Analysis of
Fork-Join Systems

Emerging Operations Research Methodologies and Applications

Series Editors: **Natarajan Gautam,** *Texas A&M, College Station, USA*
A. Ravi Ravindran, *The Pennsylvania State University, University Park, USA*

Multiple Objective Analytics for Criminal Justice Systems
Gerald W. Evans

Design and Analysis of Closed-Loop Supply Chain Networks
Subramanian Pazhani

Social Media Analytics and Practical Applications
The Change to the Competition Landscape
Subodha Kumar and Liangfei Qiu

Analysis of Fork-Join Systems
Network of Queues with Precedence Constraints
Samyukta Sethuraman

For more information about this series, please visit: https://www.routledge.com/Emerging-Operations-Research-Methodologies-and-Applications/book-series/CRCEORMA

Analysis of Fork-Join Systems

Network of Queues with Precedence Constraints

Samyukta Sethuraman

CRC Press

Taylor & Francis Group

Boca Raton   London   New York

CRC Press is an imprint of the
Taylor & Francis Group, an **informa** business

First edition published 2022
by CRC Press
6000 Broken Sound Parkway NW, Suite 300, Boca Raton, FL 33487-2742

and by CRC Press
4 Park Square, Milton Park, Abingdon, Oxon, OX14 4RN

CRC Press is an imprint of Taylor & Francis Group, LLC

ISBN: 978-0-367-71263-1 (hbk)
ISBN: 978-0-367-71264-8 (pbk)
ISBN: 978-1-003-15007-7 (ebk)

DOI: 10.1201/9781003150077

Publisher's note: This book has been prepared from camera-ready copy provided by the authors.

*To my parents, Shailaja and Sethuraman,
my doctoral thesis advisor Professor Gautam,
and my husband Ankush.*

Contents

List of Figures

List of Tables

Preface

Many real-world applications have been successfully modeled as queueing systems since the early 20th century. In the rich queueing literature that has evolved since then, a very wide range of queueing systems have been analyzed. However, many of these queueing systems continue to remain analytically intractable.

This book is about one such family of queueing systems: fork-join queues. These queueing systems are frequently encountered as parallel processing systems in a diverse set of applications including big data technologies, manufacturing systems, healthcare, and wireless sensor networks. In spite of their wide applicability, efficient estimation of performance measures of these queueing systems has remained elusive to researchers.

While working on my doctoral thesis, I ran a simulation of a fork-join queueing system. Out of curiosity, I plotted the mean response time of the fork-join system obtained from simulations, against the mean response time of a simple single-station queue that is analytically tractable. The plot turned out to be a very convincing linear line. The linearity is exciting because it opens up the possibility of estimating the performance of a complex system by estimating the performance of a simple system and exploiting the linear relationship between them.

The discovery of this linear relationship marks a significant progression in the analysis of fork-join queues and became the motivation for this work. This book is written to be a one-stop-shop for: (i) learning about different forms of fork-join queueing systems and their applications, (ii) getting up to speed on prior literature on fork-join queues, and (iii) understanding how a linear relationship between performance metrics opens up new avenues to be explored in the field of fork-join queues.

It is my hope that this book results in an increased interest and significant progress in the understanding of fork-join queueing systems in the future.

Author Biography

Samyukta Sethuraman is a Senior Research Scientist at Amazon. Her areas of expertise include queueing systems and pricing and yield management. Samyukta holds a Doctorate in Industrial Engineering from Texas A&M University and a Bachelors from Indian Institute of Technology (IIT) Madras, India. She serves on the advisory council of the Industrial & Systems Engineering Department at Texas A&M University. One of the focus areas of her doctoral thesis was fork-join queueing systems, which formed the basis of this work.

CHAPTER 1

Basic Queueing Theory

Queues are necessary elements of human civilization. How a queueing system is managed has the potential to affect how efficiently humans are able to utilize one of the most precious resources − time. Due to the importance of queueing systems in everyday life, more than half a century of research has resulted in the evolution of rich queueing theory literature. In this chapter, a brief overview of the fundamentals of queueing theory is provided. The purpose of the overview is to introduce basic concepts related to queueing systems required for a better understanding of this book.

1.1 DISCRETE SINGLE-STATION QUEUEING SYSTEMS

The queueing systems analyzed in this book are discrete queueing systems. This means that they have discrete entities entering and exiting the system. These entities are referred to as "jobs" in this book. In practice, the description of these jobs varies based on the application. For example, jobs can refer to customers in a grocery store's checkout queue or applications executing on a computer.

In this chapter, queueing theory concepts associated with single-station queueing systems are introduced. These concepts provide the necessary fundamentals for analyses of multi-station queueing systems presented in future chapters since each station in a multi-station queueing system can be analyzed as a single-station queue.

Figure 1.1 depicts a single-station queueing system. Jobs that require service arrive and join the waiting area. There are one or more servers providing service to the jobs. When a job arrives, if there is at least one free server, the service of the job starts immediately. If there is no free server, the job waits in the queue. When a server becomes available, one of the jobs waiting in the queue starts getting serviced. Note that service does not necessarily have to be according to a First Come First Served service discipline, although that is the most common service discipline analyzed in the literature. When a job's service is complete, it departs from the queueing system. With this definition of a single-station queueing system, the next section introduces a compact representation of single-station queues.

Figure 1.1: A single-station queueing system

1.2   KENDALL NOTATION

A queueing system is defined by characteristics such as the distribution of the time between the arrival of two consecutive jobs (inter-arrival time), distribution of the service time for each job, number of servers, etc. Kendall notation is a compact representation of the characteristics of a queueing system that is accepted worldwide. It is named after D.G. Kendall who proposed the first version of this notation [30]. Kendall notation consists of five components and presents as $AP/ST/NS/Cap/SD$. Each of the five components is described below:

- AP: AP denotes the queueing system's arrival process and represents the probability distribution of inter-arrival times. AP can either refer to a specific distribution or any general distribution. Examples of AP components referring to a specific distribution are M (denoting exponential), E_k (denoting Erlang-k), PH (denoting phase-type), D (denoting deterministic), etc. Alternatively, G is used to refer to any general distribution. If the arrival process is a renewal process, some authors specify this by denoting any general distribution with GI.

- ST: ST denotes the probability distribution based on which the job service times are realized. Similar to AP, ST can also refer to specific distributions or any general distribution. The examples for AP carry over to ST as well.

- NS: NS stands for the number of servers in the queueing system and is a positive integer.

- Cap: Cap stands for the capacity of the queueing system and denotes the maximum number of jobs that can exist in the queueing system at any point in time. This is an optional component of the Kendall notation with a default value of infinity.

- SD: SD refers to the service discipline based on which the decision on the next job to be served is made by the queueing system. Examples of service disciplines are First Come First Served (FCFS), Last Come First Served (LCFS), Shortest Processing Time First (SPTF), etc. This is also an optional component of the Kendall notation, with FCFS being the default value.

Under Kendall notation, $M/M/1$ represents a queueing system with exponential inter-arrival and service time distributions, one server, infinite capacity, and FCFS service discipline; while $M/G/s/K/LCFS$ represents a queueing system with exponential inter-arrival time distribution, general service time distribution, s servers, capacity K and Last Come First Served service discipline.

In the rest of this book, the queueing stations encountered fall into the category of $G/G/s$ queues with independent and identically distributed inter-arrival and service times. Therefore, the concepts presented henceforth in this chapter are applicable to queueing systems that satisfy this property. Note that these concepts *might* not necessarily be applicable to other categories of queueing systems.

1.3 QUEUEING SYSTEM PERFORMANCE

The inter-arrival and service time distributions constitute the first two components of the Kendall notation. The parameters associated with these distributions have a significant impact on the performance of the queueing system. Some of the critical parameters of inter-arrival and service time distributions needed for quantifying the performance metrics of a queueing system are listed below:

- Mean arrival rate (λ): This parameter is the average number of jobs entering the queueing system per unit time, *i.e.* λ is equal to the inverse of the mean of the inter-arrival time distribution.

- Mean service rate (μ): This parameter is the average number of jobs whose service can be completed by one server per unit time, *i.e.* μ is the inverse of the mean of the service time distribution.

- Traffic intensity (ρ): The average load experienced by the queueing system is denoted by ρ and is related to λ and μ as:

$$\rho = \frac{\lambda}{s\mu} \tag{1.1}$$

1.3.1 Flow Conservation and Stability

Queueing systems are part of a broader class of flow systems that follow the principle of flow conservation. The flow conservation principle states that at any point in time, the number of jobs that have entered the system is equal to the sum of: (i) the number of jobs present in the system at that point in time, and (ii) the number of jobs that have finished their service and left the system. In other words, jobs are neither created nor destroyed in the queueing system.

Assume that the queueing system is observed starting at time, $t = 0$. At any point in time, $t \geq 0$, assume that $A(t)$ jobs have entered the queueing system, $L(t)$ are present in the system, either waiting in queue or being served at a server, and $D(t)$ jobs have finished their service and left the system.

This flow conservation principle is expressed as Equation 1.2 below:

$$A(t) = L(t) + D(t) \tag{1.2}$$

A foundational concept associated with a queueing system's performance is the concept of stability. Rather than a mathematically rigorous definition, a more intuitive definition of stability is provided here. A stable queueing system is one in which the number of jobs present in the system is finite at any point in time, $t \geq 0$ and in particular when $t \rightarrow \infty$. This gives rise to the following definition of stability:

$$\lim_{t \to \infty} L(t) < \infty \quad almost\ surely \tag{1.3}$$

Using Equation 1.3 in Equation 1.2 gives rise to the following consequence of stability of queues:

$$\lambda = \lim_{t \to \infty} \frac{A(t)}{t} = \lim_{t \to \infty} \frac{D(t)}{t} \tag{1.4}$$

In words, Equation 1.4 simply states that in a stable queueing system, the long-run average arrival rate of jobs into the system, λ, is equal to the long-run average departure rate. The long-run arrival and departure rates, however, can be equal only if there is enough service capacity in the queueing system to be able to serve all the arriving jobs. This leads to the following condition of stability of a $G/G/s$ queue:

$$\rho = \frac{\lambda}{s\mu} \leq 1 \tag{1.5}$$

In the next section, performance metrics of queueing systems and relationships between these metrics are discussed.

1.3.2 Performance Metrics

A decision-maker evaluating a queueing system makes decisions on number of servers needed, service discipline to follow, etc. to be able to achieve a certain level of performance as defined by the performance metrics that the system demonstrates. Generally, it is hard to obtain closed-form expressions for the performance metrics if the queueing system is in transient state, *i.e.* if it is observed when the time, $t < \infty$. However, in many cases, steady-state performance metrics, *i.e.* performance metrics when $t \rightarrow \infty$, can be computed easily. Some of the most commonly used steady-state performance metrics are listed below:

- p_j: The probability that there are j, $j \in 0, 1, \ldots$ jobs in the queueing system in steady state.

- π_k (π_k^*): The probability that an arriving (departing) job observes k other jobs in the system at the time of arrival (departure).

- $F(x)$: The cumulative distribution function of the sum of the remaining service times of all the jobs in the queueing system in steady state.

- L: The expected number of jobs in the system in steady state, which is equal to $\lim_{t \to \infty} E[L(t)]$.

- T: The expected time a job spends in the system in steady state, also commonly referred to as response time or sojourn time.

The above performance metrics can be expressed in closed form for many single-station queueing systems, and the reader is referred to [25] for a detailed discussion.

1.3.3 Little's Law

Little's law establishes the following powerful relationship between the performance metrics L and T defined above:

$$L = \lambda T \tag{1.6}$$

Little's law is powerful because it is almost universally applicable to not just queueing systems but also to the more general category of discrete flow systems, irrespective of any of the components of the Kendall notation. As a result of Little's law, it becomes sufficient to know the steady-state average response time, T, to infer the average number of jobs in the system in steady state, L.

1.3.4 PASTA

The final fundamental queueing theory concept covered in this chapter is related to Poisson processes. PASTA stands for "Poisson Arrivals See Time Averages." Over a period of time, assume that a specific performance metric of a queueing system is recorded at the exact points in time when events of a Poisson process occur. According to PASTA, as the period of time approaches infinity, the average of the recorded metric is equal to the average of the metric had it been recorded continuously.

This is particularly useful in the context of queueing systems when the arrival process is Poisson, *i.e.* the inter-arrival time is exponentially distributed ($M/G/s$ queues). In these queueing systems, the steady-state performance metrics can be computed based on the value of the metric as observed by arriving jobs and don't require continuous monitoring of the metric to compute the average over time.

This concludes the chapter on queueing theory fundamentals and sets the stage for a detailed analysis of fork-join queues in forthcoming chapters.

CHAPTER 2

Introduction to Fork-Join Queues

Fork-join queueing networks are encountered in a wide range of applications. In this chapter, fork-join queues are introduced under a queueing theoretical framework. A key performance metric of these queueing systems, the mean response time, is defined; and applications are described to motivate the analysis in subsequent chapters.

2.1 INTRODUCTION

Throughout history, many advancements made by human civilizations have been geared towards doing more in less time. Parallel processing is one such advancement. A parallel processing system in its simplest form consists of a fork operation in which a job is divided into tasks that can be processed in parallel. Once all the tasks are complete, a join operation assembles all the completed tasks together to complete the original job. Quantification of the reduction in time to complete a task with parallel processing is critical for designing effective parallel processing systems.

As a simple example, consider a tech company that needs to deliver a product to a customer. Assume that the product requires the completion of two tasks: (i) developing and training a machine learning model, and (ii) software infrastructure to put the machine learning model in production. The two tasks can potentially be completed in parallel by two teams, or sequentially by one team.

Each team requires three to five weeks, with an average of 4 weeks to complete one task. Assume that the join operation of putting the machine learning model in production with the new infrastructure requires one week. The total time to deliver the product would be five weeks on an average in the parallel processing scenario, while in the sequential processing scenario, it would require an average of nine weeks. In this example, the benefit of two teams working in parallel in terms of difference in the average time to deliver is $9 - 5 = 4$ weeks. The four week difference represents a straightforward quantification of the reduction in time needed to complete a task with parallel processing.

Now consider a slightly more complex scenario, where each team has its own queue of pending projects. The teams work on these projects according to a first come first served policy. A new product gets added to the end of the existing queue. The business needs to estimate the time needed to deliver a product. The time needed to deliver the product is equal to the sum of: (i) the maximum of the time needed by each of the two teams to work through their respective queues and complete their task, and (ii) time needed to complete the join operation. In this scenario, estimating the time to deliver the product (also referred to as the response time) is not straightforward since the time required by each team to complete the tasks in their queues is stochastic, and the number of tasks in the two queues are correlated.

Analysis of fork-join queues helps to estimate the time needed to complete jobs being processed in parallel in queueing systems. This is a critical step towards enabling decision-makers in a wide variety of applications to make informed and data-driven decisions.

The next section gives a queueing theory-based definition of response time in fork-join queues.

2.2 RESPONSE TIME IN FORK-JOIN QUEUES

In a fork-join queueing network, jobs arrive at a fork node, denoted by "F" in Figure 2.1, at random time intervals sampled from a known distribution. At the fork node, these jobs get partitioned into multiple tasks. Each task itself might consist of multiple sub-tasks in series, parallel, or a combination of series and parallel, with their individual sub-task queue. An arriving job undergoes service at each of the n tasks, denoted by $1, 2, \ldots, n$ in Figure 2.1. The service time of a task or sub-task is random with a known distribution. Due to the stochasticity introduced by random service times, it is possible and likely that one task of a job is complete but another task is not. After the completion of a task, the task waits in a buffer for other tasks belonging to the same job to complete. Based on the system specification, either some or all of the tasks have to be complete for the job to make it to the join node, denoted by "J" in Figure 2.1. Once all necessary tasks are complete, the tasks are joined at the join node to complete the job before it exits the system.

A key performance metric of interest in this fork-join queueing network is the average time spent in the system by a job in steady state, known as the mean steady-state response time of the system. Exact estimation of this quantity has been acknowledged to be notoriously difficult [33,69]. The difficulty arises from the dependence caused by the synchronized arrivals and departures from the queueing system. In this book, some new insights into the properties of the response time in these queueing systems are presented. These insights lead to the formulation of algorithms to approximate the mean response time of these systems to a significantly higher degree of accuracy than previously observed in the literature.

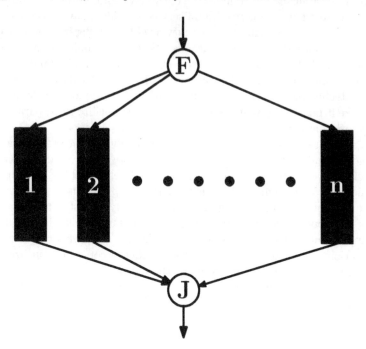

Figure 2.1: A fork-join queueing network

2.3 APPLICATIONS

Fork-join queueing networks are encountered in a diverse set of applications ranging from manufacturing systems to health care to wireless sensor networks. Some of these applications are described in detail in this section.

2.3.1 Parallel Computing

Interest in fork-join queueing networks arose in the 1980s when the first parallel processing computer systems were prototyped. Lavenberg [37], in his work on fork-join queueing networks in 1989, states: "While parallel processing of a job is possible on current computer systems, it is expected to become much more prevalent in the future," and his prediction has definitely proved to be accurate. In the year 2001, IBM introduced the first commercially available dual-core computer system with parallel processing capability [19]. The first android phone with two cores was introduced by LG in 2011 [1]. Currently, commercially available desktop computer systems can have upto 28 cores, and commonly used laptop computers have 6-8 cores. Cloud service providers have taken parallel processing capabilities to a new level by making it possible to utilize hundreds or even thousands of processing units running in parallel. Software capabilities have also evolved hand in hand with the evolution of parallel processing hardware. Big data processing technologies such as Apache Spark and Hadoop MapReduce make it possible to write simple code that can utilize these parallel computing capabilities.

In parallel computing systems, a job is divided into tasks that can be independently processed by each core or server in a cluster. Consider the example of a random forest machine learning model [15] implemented on a regular laptop computer with six cores. The random forest model can potentially consist of hundreds of independent decision trees. The commonly used Scikit-learn [49] implementation of the random forest algorithm allows for the utilization of multiple cores, where the number of cores to be utilized is given as an input parameter. If the option to use all available cores is chosen, in a 6-core laptop, the algorithm trains six trees in parallel until the required number of trees is reached. Finally, results from all the trees are aggregated in a join operation.

Other examples of parallel computing are parallel data retrieval from multiple storage disks and query processing in distributed databases [17, 29, 44, 59, 72]. The division of a large job into tasks, distribution of the tasks among processors, and final aggregation of results from individual tasks naturally gives rise to the formation of a fork-join queue. Estimation of the response time in these systems using analysis of fork-join queueing networks enables efficient decision making on the number of parallel processors to use and the service rate required from each processor.

2.3.2 Manufacturing Systems

Another established application of a fork-join queueing network in literature is parallel assembly lines in a manufacturing system. The following simple example from the work by Nguyen [48] explains how fork-join queues are encountered in assembly lines. Consider a shop that manufactures coats. An arrival of a job into the system corresponds to an order for a coat from a customer. The job of stitching this coat consists of tasks that can be processed in parallel. For example, if the shop is adequately staffed, the sleeves, fronts, backs, collars, pockets, lining, etc. could be prepared at the same time. Once all the above are ready, the coat is ready for the final assembly, which corresponds to the join process. After the final assembly, the coat is ready to be delivered to the customer, which corresponds to the job leaving the system. This is a simple example of a fork-join queueing network in a manufacturing facility. More complex networks involving multiple sub-tasks within one task might be necessary for modeling more sophisticated processes such as car assembly. If the product requires the same server to process multiple tasks, this can be modeled as a multi-class fork-join queueing network [28].

2.3.3 Health Care Systems

Fork-join queues are also encountered in multiple health care applications. Consider the process of a physician arriving at a diagnosis. A patient entering a hospital for consultation corresponds to arrivals to the system. The physician might need to prescribe a set of diagnostic tests which can be carried out in parallel. The processing of each of the test samples would possibly be conducted by different departments, each with its own queue. Each test corresponds to a parallel task in a fork-join queue. Once all the test results are available, the physician arrives at a diagnosis which corresponds to the job exiting the system after completing service in a fork-join system. In another application, Carmeli et al [16] model announcements in an emergency department as fork-join queues. The authors also demonstrate the fork-join structure of patient flow as an animation in [57].

In a different setting, processes taking place in the human brain have been modeled as fork-join queueing networks by Liu [41]. Stimulants that require a response from the human body are modeled as jobs entering the queueing system. The author points to instances of the brain processing different aspects associated with the stimulant in parallel. Each aspect is a task in the fork-join queue. The response time in this scenario is the time between when the stimulant is encountered and when the human body responds to it.

2.3.4 Wireless Sensor Networks

Fork-join queues are also encountered in wireless sensor networks set up in cities such as Singapore [27]. Sensors capable of sensing environmental factors such as temperature, humidity, air pollution, and noise level are installed on lamp posts in the city. The controlling authority is interested in a function of the data being collected, for example the maximum temperature. All the sensors sense at the same instants of time. The time interval between consecutive sensing instants of time is random with a known distribution. This sensing of data corresponds to arrivals to the fork-join queueing network. Once a set of data has been sensed, each sensor transmits it to the controlling authority. The sensors can transmit only one data point at a time. It takes a random amount of time with a known distribution to complete the transmission of one data point from a sensor to the controlling authority. This corresponds to the service time of a task in the fork-join queueing network. If a set of data is sensed while transmission is taking place at a sensor, the new data point waits in a queue at the sensor for its turn to be transmitted. For each set of data that has been sensed at the same time, the controlling authority waits to receive data points from each sensor before computing the function of the data. Computing the function takes a relatively small amount of time since the functions being computed, such as maximum and mean, are simple. Once the required functions of a set of data have been computed, the function values are recorded, and this corresponds to the departure of the job from the system.

Fork join queues arise in wireless sensor networks used in many other settings [13]. For example, wireless sensors can be used to monitor water and electricity utilization in smart cities. In these networks, computation of the net utilization based on data measured by each sensor leads to the formation of a fork-join queue similar to that in the example of wireless sensors on lamp posts described above.

In environmental applications, wireless sensors are installed at locations that are hard to reach and can be used for monitoring wildlife habitats and forest fires. Forest fires, for example, can be detected by computing the maximum of temperatures recorded by sensors in an area. Computation of this maximum temperature can again be modeled as a fork-join queue as in the previous two examples. Efficient design of wireless sensor networks based on analysis of fork-join queues can enable faster response to wildfires.

2.4 ORGANIZATION

The rest of this book is organized as follows: In Chapter 3, a comprehensive literature review on fork-join queues is presented. Following that, in Chapter 4, a new and efficient algorithm is proposed to estimate the response time of the most well-studied type of fork-join queues − the symmetric n-dimensional fork-join queueing system. Finally, in Chapter 5, this algorithm is extended to more complex forms of fork-join queues for which there are no response time approximations available in the literature.

CHAPTER 3

Literature Review

Research on fork-join queueing networks was first seen in literature in the 1980s, matching the time-line of the first parallel processing computer systems. Since then multiple researchers have focused their attention on these queueing systems and contributed to breakthroughs in their understanding. In this chapter, literature on this subject is organized into two sections. In the first section, a review of the literature on fork-join queues with one level of parallel tasks is presented. More complex network structures are reviewed in the second section.

The majority of the work in literature assumes symmetry of tasks, FCFS service discipline, and single server queues in the fork-join queueing network. Symmetry of tasks means that each individual task queue in the network follows the same service time distribution. In this chapter, unless specifically mentioned, symmetry of tasks, FCFS service discipline, and single server task queues are assumed in the reviewed literature.

3.1 FORK-JOIN QUEUEING SYSTEMS WITH A SINGLE LEVEL OF TASKS

Figure 3.1 depicts a fork-join queueing system with a single level of parallel tasks. An arriving job instantaneously forks into n, $n > 1$ parallel tasks, each with its own queue. Based on the system design, either some or all the tasks have to complete for the job to be considered complete. The default system in the literature reviewed in this section, unless specified otherwise, is one in which

DOI: 10.1201/9781003150077-3

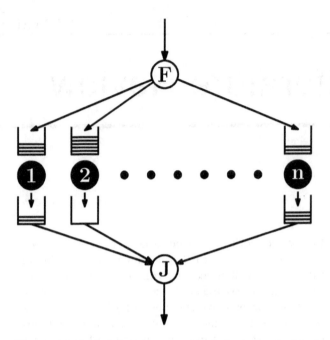

Figure 3.1: A fork-join queue with a single level of tasks

all the n tasks have to complete for the job to be complete. Once all necessary tasks are complete, the job moves to the join node and leaves the system.

One of the first advances in response time estimation of fork-join queues is the work by Flatto and Hahn [24]. The authors analyze a fork-join queueing system with two not necessarily symmetric parallel tasks and exponential inter-arrival and service time distributions. They derive the probability generating function of the bi-variate distribution of the number of tasks in the two task queues. Using this result, they also derive the asymptotic joint probability distribution of the number of tasks in the two queues as the number of tasks in one queue approaches infinity. Following up on this work, Flatto [23] analyzes the asymptotic interdependence between the number of tasks in the two queues as these numbers approach infinity.

Flatto and Hahn's [24] result was used by Nelson and Tantawi [45] to derive the closed-form expression of the mean response time in steady state in a fork-join queueing system with two parallel tasks and exponential inter-arrival and service time distributions.

To this date, this remains the only closed-form exact result on the response time of a fork-join queueing system. Nelson and Tantawi [45] also propose an approximation for the mean response time of a fork-join queueing system with more than two parallel tasks. This approximation is based on an interpolation between upper and lower bounds to the mean response time. The authors construct the interpolation approximation based on the observation that both the upper and lower bounds increase linearly with respect to the harmonic sum of the number of tasks.

Flatto and Hahn's [23] analysis technique was extended to the case of a two-task fork-join queueing system with exponential inter-arrival and general service time distributions by Baccelli [3]. The author proved that the transient and steady-state workload Laplace transforms can be obtained as the solution to boundary value problems.

Varma and Makowski [70] provided a closed-form approximation for the mean response time in steady state in fork-join queueing systems with general inter-arrival and service times. This approximation is computed as an interpolation between light and heavy traffic approximations of the response time. The light traffic approximation is obtained using the light traffic interpolation technique for open queueing systems proposed by Reiman and Simon [52]. The heavy traffic limit is computed as an extension of the diffusion limit of a $G/G/1$ queue.

Thomasian and Tantawi [60] propose approximations for the mean steady-state response time in fork-join queueing systems with exponential inter-arrival times and general service times. This approximation requires the computation of parameters using multiple simulations. The authors demonstrate the higher accuracy of this approximation when compared against the approximation by Varma and Makowski [70] for larger values of the number of parallel tasks.

Ko and Serfozo [33] also proposed a closed-form approximation for the steady-state response time distribution in a fork-join queueing system with exponential inter-arrival and service times, when the number of servers in each task node can be greater than one. This approximation does not require the symmetry of task queues. In another work by Ko and Serfozo [34], the authors extend the analysis to provide closed-form approximations for the distribution of the steady-state response time in fork-join queueing systems with general inter-arrival times and exponential service times.

Varki and Dowdy [66] use a technique called event/time sequence trees to analyze properties of the mean response time in a fork-join queue with two tasks, exponential inter-arrival times and negative exponential service times. The authors prove that the parallel service time (defined as the time between when the service of the last task starts and when the job exits the system) is dependent on the state of the queueing system at the time when the job arrives into the system. This property makes the analysis of fork-join queueing systems notoriously hard. Varki [65] proposes an iterative numerical technique to estimate the mean response times in steady state of fork-join queueing systems. The inter-arrival and service time distributions are assumed to be exponential. In another work, Varki et al [67] propose a closed-form approximation formulated as the mean of lower and upper bounds of the mean response time of (n, k) fork-join queueing systems. An (n, k) fork-join queue has n task queues. An arriving job, however, needs only k complete tasks, where $k \leq n$, to be able to exit the system. How the k tasks are chosen out of n is modeled differently by different authors.

Lebrecht and Knottenbelt [38] compare the performance of the approximations of the mean response times proposed by Nelson and Tantawi [45], Varma and Makowski [70], and Varki et al [67] with the upper bound computed by assuming that the service of a job starts only when all the sub-tasks of the previous sub-tasks are complete. They demonstrate higher accuracy of the approximations by Nelson and Tantawi [45] and Varma and Makowski [70] compared to the approximation by Varki et al [67] and the upper bound, for fork-join queueing systems with exponential inter-arrival and service time distributions. The authors further present results to demonstrate the performance of the upper bound against simulations for fork-join queueing systems with non-exponential service times and non-symmetric tasks.

Kemper and Mandjes [31] propose an approximation for a two-task fork-join queue with exponential arrival and general service time distributions. This approximation is formulated as a linear function of the first two moments of the squared coefficient of variation of the service time distribution. Khabarov et al [32] propose an approximation for the mean response time in steady state for a fork-join queueing system with exponential inter-arrival and hyper-exponential service time distributions. This requires the approximation of the steady-state response time of $M/G/n$ queues (single-station queues with n servers, where n is the number of tasks in the original fork-join queue) using iterative techniques. The authors demonstrate the higher accuracy of this approximation when compared with the approximations by Nelson and Tantawi [45], Varma and Makowski [70], and Varki et al [67] at higher traffic intensities and higher number of tasks.

Qiu et al [51] use matrix analytic methods to derive the distribution of the steady-state response time in a fork-join queueing system with phase-type service time distribution and Markovian arrival process. The authors analyze the system under constraints on the difference in the number of tasks in the longest and shortest task queues. This analytic technique becomes intractable when the number of tasks is greater than two. For fork-join queues with number of tasks greater than two, the authors propose an approximation for response time tails based on order statistics.

Nelson et al [46] introduce a form of (n, k) fork-join queues where the number of tasks to be completed, k, is a discrete-valued random variable with a known probability mass distribution. Under assumptions of exponential inter-arrival and service time distributions, the authors propose an iterative solution to approximate the mean response time. Towsley et al [61] extend this work to the case where the servers follow a processor sharing service discipline. They derive bounds, propose approximations, and compare the processor sharing service discipline against FCFS service discipline. Their findings show that systems operating under FCFS are more efficient in terms of response times. Varki and Zhang [68] extend the iterative technique in [65] to this form of fork-join queues. For these systems, the authors compute simple upper and lower bounds to the mean response time.

Thus far into this chapter, a review of major advances in the approximation of the response time in fork-join queueing systems is presented. In Chapter 4, a new closed-form approximation for the mean response time is proposed and compared with applicable closed-form approximations from the literature.

Although many approximations have been proposed, there are no exact results for the response time of fork-join queues when the number of tasks, $n > 2$. Due to this lack of exact results, there is a plethora of literature on bounds on the response time. Baccelli and Makowsky [6] analyze a fork-join queueing system with general renewal sequences of inter-arrival and service times. They compute upper and lower bounds on response times by bounding the response time of the fork-join queue with the waiting times of more tractable queueing systems with the same stability condition as the original fork-join queue. The lower bound is the response time in a fork-join queue with deterministic arrivals, while the upper bound is the response time of a fork-join queue with non-synchronized and independent arrivals. These bounds hold true for both steady-state and non-steady-state conditions, and even for fork-join queues with non-symmetric tasks. Similar bounding concepts are also used by Baccelli and Liu [5] to compute bounds for fork-join queueing systems in which multiple tasks are assigned to the same server.

Balsamo and Mura [11] analyze a fork-join queueing system with non-symmetric tasks and exponential inter-arrival and service time distributions. The authors compute upper and lower bounds to the response time distribution under both transient and steady-state conditions using matrix-geometric analysis of Markov chains. They extend this analysis to fork-join queues with Coxian and phase-type arrival and service time distributions in [12] and [10].

Lui et al [43] compute bounds on the mean response time in steady state in a fork-join queueing system with phase-type inter-arrival time and Erlang service time distributions. They achieve this by modifying the state space of the underlying stochastic process to more mathematically tractable forms that serve as lower and upper bounds on the response time distribution.

Similar to the fork-join queueing system, which forms the lower bound in the system analyzed by Baccelli and Makowsky [6], Pinotsi and Zazanis [50] also analyze a fork-join queueing system with deterministic arrivals and exponential service times. The reason for this system being tractable is the independence between queues in the Palm version [4] induced by the deterministic arrivals. In this work, the authors obtain closed-form expressions for the joint probability distribution of the steady-state number of entities in each of the n queues. Rizk et al [53] also extend the work of Baccelli and Makowsky [6] to include renewal as well as non-renewal arrival streams. The authors compute bounds on the steady state response time using results from martingale theory and demonstrate the tightness of bounds using simulations.

Joshi et al [29] model a distributed disk storage system as an (n, k) fork-join queue with exponential inter-arrival and service time distributions. The authors derive the stability condition of this system and compute bounds on the mean response time in steady state. They use these bounds to demonstrate the trade-off between the number of sub-tasks k and the mean response time.

Apart from analytic techniques, simulation of queueing systems is another approach to estimate the performance measures of queueing systems. Since simulations require significant time and computing resources, Dai [20] and Chen et al [18] design efficient simulation techniques for estimating performance measures of fork-join queueing systems.

In the next section, a review of literature on fork-join queueing networks with a more general network structure is presented.

3.2 GENERALIZED FORK-JOIN QUEUEING SYSTEMS

Many researchers have analyzed fork-join queueing networks with different assumptions on the network structure. One of the first works is by Baccelli and Massey [8]. The authors analyze queueing networks that are a combination of regular queues in series and fork-join queues, with the output stream of one queue acting as the arrival stream to another. The authors utilize the concepts related to queues with higher and lower variability from their previous work [6]. They combine these with results on convex ordering to obtain upper and lower bounds on the response time in both transient and steady states. In Baccelli et al [9], the authors introduce acyclic fork-join queueing networks. In these queueing networks, there is a single arrival stream and departure stream. The nodes in the network follow pre-set precedence constraints. These precedence constraints result in an acyclic queueing network with multiple fork and join nodes. The authors use similar bounding techniques as in [6] and [8] along with results on associated random variables to obtain bounds on the response time in both steady and transient states. Properties of associated random variables are also used by Kumar and Shorey [36] to derive bounds for fork-join queueing networks with multiple single level (n, k) fork-join queueing systems in series. The authors also propose lower bounds based on more tractable queueing systems defined by neglecting queueing delays or assuming deterministic arrivals.

Ammar and Gershwin [2] analyze a fork-join queueing network with limited buffer capacities and blocking. The authors show equivalence relationships between fork-join networks with different precedence constraints constructed by changing the traffic flow direction of a subset of the nodes. Gershwin [26] extends this work to approximate the throughput of an acyclic fork-join queueing network with limited buffer space and blocking. The author approximates the throughput of the system by decomposing the network into smaller networks. Dallery et al [21] prove reversibility properties of fork-join queueing networks with limited buffer capacities and blocking. The authors prove that the throughput of such a network is a concave function of the buffer sizes and the initial number of tasks in each buffer. These properties are used to construct throughput optimal networks.

Another generalization of a fork-join queueing system is a queueing system with both synchronized and non-synchronized arrival streams. Wright [71] analyzes a fork-join queue with two tasks and exponential inter-arrival and service times. Apart from the fork-join arrival stream, the two task queues also serve arrivals from another non-fork-join arrival stream with exponential inter-arrival time distribution. The author uses a technique similar to Flatto and Hahn [24] to obtain the probability generating function of the bi-variate distribution of the number of tasks in the two queues. Towsley and Yu [62] analyze a similar two-task fork-join queueing system in which, along with the forking jobs, there is another independent arrival stream of jobs that are served at only one task queue. The authors assume exponential inter-arrival and service time distributions. Under these assumptions, the authors compute bounds on the response time distribution in steady state. Guide Jr. et al [28] present an analysis of a similar queueing system in the context of a re-manufacturing facility. The authors provide an approximation for the average weighted job response time by approximating the correlated waiting times at the two task queues as independent waiting times.

Multiple other generalizations of fork-join queueing systems have been modeled by researchers. For example, closed fork-join queueing networks have been analyzed by Duda and Czachórski [22], Liu and Perros [42], Lee and Katz [39], Varki [64], Varki [65] and Krishnamurthy et al [35]. Squillante et al [54] introduce a generalization of the fork-join queueing network in which each task consists of different stages of execution. The network model also incorporates task vacations and communication among tasks. The authors present a matrix-analytic analysis of this generalized fork-join queueing system. Takahashi et al [55] and Takahashi and Takahashi [56] analyze parallel server queueing systems where arrivals to the servers are independent and not synchronized, but synchronization happens at the time of departure from each queue. On the other hand, Li and Xu [40] compute bounds on the joint distribution of the number of entities in each queue in a parallel server system where arrivals are synchronized, but departures are not. The joint distribution in this system is the same as the joint distribution of queue lengths in a fork-join queue. Nguyen et al [47] introduce a fork-join queueing system where each task node itself can be a fork-join queue. The authors provide approximations for the response time under heavy traffic conditions and general inter-arrival and service times.

Response times or queue lengths have received the most attention in terms of performance measures of fork-join queueing networks. However, some researchers have also analyzed other performance measures. For example, Tsimashenka and Knottenbelt [63] construct an online algorithm to delay the start of service at some tasks to minimize the time between completion of the first and last tasks in a fork-join queueing system with general inter-arrival times, exponential service times and heterogeneous servers. Xia et al [73] derive necessary conditions for the throughput scalability of a finite buffer acyclic fork-join queueing network and evaluate the scalability as the number of tasks approaches infinity. Zeng [74] extends this analysis to derive necessary as well as sufficient conditions for throughput scalability as the number of tasks approaches infinity. Zhao et al in [75] and [76] analyze efficient resource allocation policies in fork-join networks. Carmeli et al [16] model announcements in an emergency department as an application of fork-join queues. The authors propose an exact analysis technique to estimate the delay distributions in fork-join queues in transient state using recursive Laplace Stieltjes transforms of the joint distributions. For comprehensive reviews on systems with parallel tasks, the reader is referred to the reviews by Boxma et al [14] and Thomasian [58]. With this understanding of the literature on fork-join queueing systems, in the next chapters a rigorous queueing theory based analysis of these systems and their response time is presented.

CHAPTER 4

Symmetric n-Dimensional Fork-Join Queues

In this chapter, a linear relationship is conjectured between the mean response time of the symmetric n-dimensional fork-join queueing system and the mean queue length of any one task in the system. This linear relationship leads to formulation of an algorithm to estimate the mean response time of the system. Comparisons against simulations demonstrate that the mean response time eastimted using this algorithm is up to 44% closer to the estimate from simulations than existing approximations in the literature.

Symmetric n-dimensional fork-join queues are the most well-studied types of fork-join queueing networks. In these queueing systems, arriving jobs are partitioned into n stochastically identical tasks. All the n tasks need to finish service before the job can exit the system. In Section 4.1, the symmetric n-dimensional fork-join queueing system is defined under a queueing theory framework. Section 4.2 presents the main conjecture that leads to the formulation of an algorithm to estimate the mean response time of this system. In Section 4.3, extensive comparisons against simulations are presented to evaluate the performance of the algorithm against those existing in prior literature. Finally, Section 4.4 summarizes and concludes the chapter.

DOI: 10.1201/9781003150077-4

4.1 SYSTEM DEFINITION

In the symmetric n-dimensional fork-join queueing system, jobs arrive according to a Poisson process with rate λ. On arrival, each job instantaneously forks into n tasks. There are n single server queueing stations with infinite capacity and operating under an FCFS service discipline. A task with index i, $1 \leq i \leq n$, is routed to the queue at queueing station i. The n queueing stations are symmetric (*i.e.* all servers and tasks are stochastically identical) in the fork-join queueing system. The task service times are independent and identically distributed across jobs and across constituting tasks of the same job. These service times are $\mathcal{R}_+$ valued and follow a general distribution with cumulative distribution function $G(x)$ that has a finite mean and variance. The mean of $G(x)$ is denoted by $1/\mu$, *i.e.* μ is the average service rate. The traffic intensity at each queueing station is denoted by $\rho = \frac{\lambda}{\mu}$. Each of the n task queueing stations has a join buffer. After completion of a task, it is routed to the join buffer, where the task waits for the rest of the tasks belonging to it's parent job to finish service. Joining is assumed to occur instantaneously. Therefore, the departure time of the last task of a job to finish service coincides with the departure time of the job from the system. This network model is shown in Figure 4.1.

Baccelli and Makowski [6] derived the stability condition for this system, which is same as the stability condition of a $G/G/1$ queue. The symmetric n-dimensional fork-join queueing system is stable *iff* $\lambda < \mu$, which is same as $\rho < 1$.

Using index j, $j \geq 1$, for the arriving jobs in the order of their arrival, the sequence of inter-arrival times of the jobs is denoted by $\{\tau_j, j = 1, 2, \ldots\}$ and the sequence of service times at task i, $1 \leq i \leq n$, by $\{\sigma_j^{(i)}, j = 1, 2, \ldots\}$. $W_0^{(i)}$, denotes the initial workload at the queueing station of task i. The workloads observed by subsequent jobs are denoted by $W_j^{(i)}$. These are defined by the Lindley recursions as follows:

$$W_j^{(i)} = \left[W_{j-1}^{(i)} + \sigma_{j-1}^{(i)} - \tau_j \right]^+, \qquad i = 1, \ldots, n; \; j = 1, 2, \ldots$$

The time interval between the arrival of job j and its completion of task i is denoted by $T_j^{(i)}$ which is defined below:

$$T_j^{(i)} = W_j^{(i)} + \sigma_j^{(i)}, \qquad i = 1, \ldots, n; \; j = 1, 2, \ldots$$

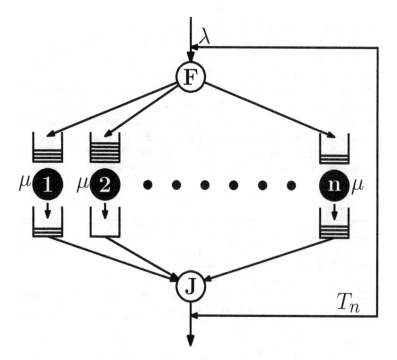

Figure 4.1: A symmetric n-dimensional fork-join queueing system

The response time of job j, denoted by $T_{j,n}$ is defined as:

$$T_{j,n} = \max_{i=1,\ldots,n} T_j^{(i)} \qquad j = 1, 2, \ldots \qquad (4.1)$$

The response time of any job in steady state is denoted by T_n. When $\rho < 1$, the steady-state distribution of T_n exists, has a finite mean represented by $E[T_n]$, and the distribution is independent of the initial state of the system. This mean steady-state response time, $E[T_n]$, is the primary system performance measure that is of interest in this analysis presented in this book. It is to be noted that the assumption of instantaneous joining of tasks at the join node does not impact the analysis. If the service time at the join node is non-zero, the response time of the system can be obtained by adding the average time spent at the join node to the estimated value of $E[T_n]$.

From Equation (4.1) it can be inferred that T_n is the maximum of n identically distributed random variables. Each of these n random variables is the response time of an $M/G/1$ queueing system whose steady-state distribution is known. However, there are no known methods for the evaluation of $E[T_n]$ using Equation (4.1) because these n random variables are not independent. The synchronized arrivals to all the n task queues introduce a correlation between the queue lengths. This correlation has been found to be extremely hard to quantify. Therefore, obtaining a closed-form expression for $E[T_n]$ has proved elusive for many researchers. In the next section, an estimation technique for $E[T_n]$ based on a conjecture is presented. In the rest of this chapter, for brevity $E[T_n]$ is denoted by $\bar{T}_n$. Table 4.1 lists the notations used.

Table 4.1: List of notations

Notation	Description
λ	Mean arrival rate
μ	Mean service rate of each task
$G(x)$	CDF of the service time distribution
ρ	Traffic intensity $(= \frac{\lambda}{\mu})$
T_n	Random variable representing the steady-state response time
$\bar{T}_n$	Expected steady-state response time

4.2 RESPONSE TIME ESTIMATION

In this section, an algorithm to approximate $\bar{T}_n$ is presented. The following conjecture forms the basis of this algorithm:

Conjecture 4.1 *The mean response time of the symmetric n-dimensional fork-join queueing system is linear with respect to $\frac{\rho}{1-\rho}$, i.e. the slope and intercept are independent of the traffic intensity, ρ. The intercept is equal to the expected value of the maximum of n i.i.d. random variables with cumulative distribution function $G(x)$.*

$$\bar{T}_n = \frac{\rho m_n}{\mu(1-\rho)} + \int_0^\infty nx\{G(x)\}^{n-1}dG(x) \qquad (4.2)$$

where, m_n is a parameter independent of ρ.

The intercept term in Equation (4.2) is explained by the following: As ρ approaches zero from the right, an arriving job sees an empty system with probability approaching one *i.e.* an arriving entity will find an empty system almost surely. Therefore, with probability approaching one, the response time of any job arriving into the system is the maximum of the service times at the individual tasks. The service times at the individual tasks are independent with cumulative distribution function $G(x)$. Hence the intercept term in Equation (4.2) is the expected value of the maximum of the n *i.i.d.* random variables representing the service time at the n individual tasks.

Remark 4.1 *Conjecture 4.1 is satisfied by systems with general service times when $n = 1$, and by systems with exponential service times when $n = 2$. For these systems, the closed-form expressions for the average response time are known. When $n = 1$, the system is an $M/G/1$ queue. The expected response time in steady state is $\frac{(1+C_v^2)\rho}{2\mu(1-\rho)} + \frac{1}{\mu}$. In this case, the parameter $m_1 = \frac{1+C_v^2}{2}$, where C_v^2 is the squared coefficient of variation of the service time distribution $G(x)$. When $n = 2$ and service times are exponentially distributed, Nelson and Tantawi [45] show that the expected response time in steady state is $\frac{12-\rho}{8\mu(1-\rho)}$ which can be written as $\frac{11\rho}{8\mu(1-\rho)} + \frac{3}{2\mu}$. In this case $m_2 = \frac{11}{8}$ and the expected value of the maximum of two independent exponentially distributed random variables with mean $\frac{1}{\mu}$ can be calculated as $\int_0^\infty 2x(1 - e^{-\mu x})d(1 - e^{-\mu x}) = \frac{3}{2\mu}$.*

Intuition behind Conjecture 4.1. The expected response time of the symmetric n-dimensional fork-join queueing system can be written in the following form:

$$E[T_n] = E[T_{n-1}] + E[T^{(n)} - T_{n-1}|T^{(n)} > T_{n-1}]P(T^{(n)} > T_{n-1})$$
$$(4.3)$$

The traditional representation of expectations is used in Equation 4.3 for clarity. $T^{(n)}$ denotes the steady-state response time random variable of the nth task. Equation 4.3 states that the expected steady-state response time of a symmetric n-dimensional fork-join queueing system, $E[T_n]$, exceeds the expected steady-state response time of a symmetric $n - 1$-dimensional fork-join queueing system, $E[T_{n-1}]$ only if the nth task, is the last one to finish. The probability of this event is represented by $P(T^{(n)} > T_{n-1})$. The excess time needed to complete the nth task, given that it is the last one to finish service, is represented by $E[T^{(n)} - T_{n-1}|T^{(n)} > T_{n-1}]$. Due to the symmetry of the tasks in the system: $P(T^{(n)} > T_{n-1}) = \frac{1}{n}$.

As discussed in Remark 4.1, Conjecture 4.1 holds when the number of tasks $n = 1$. For the purpose of using an induction argument, assume that Conjecture 4.1 holds when the number of tasks is $n - 1$. Substituting for $P(T^{(n)} > T_{n-1})$ and $E[T_{n-1}]$ in Equation 4.3 results in:

$$E[T_n] = \frac{\rho m_{n-1}}{\mu(1 - \rho)} +$$
$$\int_0^\infty (n - 1)x\{G(x)\}^{n-2}dG(x) + \frac{E[T^{(n)} - T_{n-1}|T^{(n)} > T_{n-1}]}{n}$$
$$(4.4)$$

It can be inferred from Equation 4.4 that Conjecture 4.1 holds true if $E[T^{(n)} - T_{n-1}|T^{(n)} > T_{n-1}]$ increases linearly with respect to $\frac{\rho}{1-\rho}$. The expected workloads of each of the individual $M/G/1$ queues are known to be linear with respect to $\frac{\rho}{1-\rho}$. As ρ increases, $\frac{\rho}{1-\rho}$ increases and the variance of the individual workloads increases. Therefore, it is expected that the excess time needed to complete the last task also increases with increase in $\frac{\rho}{1-\rho}$. When these quantities are plotted, they display a very convincing linear relationship.

To demonstrate the linear relationship with an example, the average response time obtained using simulations is plotted against $\frac{\rho}{1-\rho}$ for the case of $n = 5$, $\mu = 1$ and exponential service time distribution. The linear relationship is apparent in Figure 4.2.

When the task service times are exponentially distributed, the parameter m_n in Equation 4.2 displays some special properties, which are explained in the next subsection.

4.2.1 Exponential Service Time Distribution

When the service times are exponential, $G(x) = 1 - e^{-\mu x}$. In this case, $\int_0^\infty nx\{G(x)\}^{n-1}dG(x) = \frac{H_n}{\mu}$, where $H_n = 1 + \frac{1}{2} + \ldots + \frac{1}{n}$, is the harmonic sum. Furthermore, based on simulations, the parameter m_n is observed to satisfy:

$$m_n = \sum_{k=1}^n \frac{k + 1}{2k^2} \qquad (4.5)$$

As mentioned in Remark 4.1, the values of m_n for $n = 1$ and $n = 2$ are known. It can be easily confirmed that these values satisfy Equation 4.5. In Figure 4.3, values of m_n obtained using simulations are plotted against $\sum_{k=1}^n \frac{k+1}{2k^2}$. The plot demonstrates a linear trend line with slope ≈ 1 and intercept equal to zero.

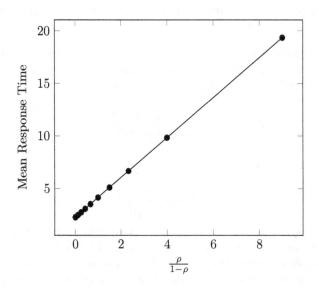

Figure 4.2: Plot of simulated average response time vs. $\frac{\rho}{1-\rho}$ for $n = 5$, $\mu = 1$ and exponential service time distribution

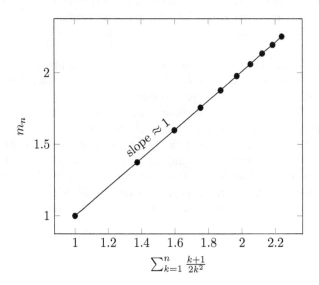

Figure 4.3: Plot of parameter m_n vs. $\sum_{k=1}^{n} \frac{k+1}{2k^2}$

Based on this observation, the conjecture for exponential service times can be written as:

Conjecture 4.2 *The mean response time of the symmetric n-dimensional fork-join queueing system with exponential inter-arrival and service time distributions is given by:*

$$\bar{T}_n = \frac{\rho}{\mu(1-\rho)} \sum_{k=1}^{n} \frac{k+1}{2k^2} + \frac{H_n}{\mu} \tag{4.6}$$

Equation 4.6 can also be written as:

$$\bar{T}_n = \bar{T}_{n-1} + \left[\frac{1+1/n}{2} \cdot \frac{\rho}{\mu(1-\rho)} + \frac{1}{\mu} \right] \cdot \frac{1}{n} \tag{4.7}$$

Interpretation of Conjecture 4.2. Comparing Equations 4.7 and 4.4, it can be inferred that $E[T^{(n)} - T_{n-1}|T^{(n)} > T_{n-1}]$, which is the expected excess time needed to complete the nth task, given that it is the last one to finish service, is conjectured to be equal to $\left[\frac{1+1/n}{2} \cdot \frac{\rho}{\mu(1-\rho)} + \frac{1}{\mu} \right]$. This can be interpreted as follows: After $n-1$ tasks are complete in mean response time $\bar{T}_{n-1}$, the job can be thought of as joining a synchronization queue with probability $\frac{1}{n}$. It is conjectured that the synchronization queue has the same mean response time as an $M/G/1$ queue with Erlang-n service time distribution and mean service rate equal to μ.

Even though Conjectures 4.1 and 4.2 have not been proved, extensive experiments establish their utility for approximating the expected response time of the symmetric n-dimensional fork-join queueing system. In the next section, an algorithm is formulated using Conjecture 4.1 to estimate the mean response time when service times are not exponentially distributed.

In the following sections, the mean response time estimate derived using Conjectures 4.1 and 4.2 is referred to as the n-1 *linear approximation*. The name is derived from the fact that the mean response time of a fork-join queue with n tasks is conjectured to be a linear function of the mean queue length of any one task in the system.

4.2.2 General Service Time Distribution

In a symmetric n-dimensional fork-join queueing system, given the number of parallel tasks n, the mean service rate μ, and the service time CDF $G(x)$, $\bar{T}_n$ only depends on the traffic intensity

ρ. *For clarity, this relationship is denoted by writing* $\bar{T}_n$ *as* $\bar{T}_n(\rho)$. When the task service times are exponentially distributed, the estimate for the mean response time is obtained directly from Equation 4.6. When the task service times are not exponential, to estimate $\bar{T}_n(\rho)$ for any traffic intensity ρ using Equation (4.2), the value of m_n needs to be estimated. To estimate m_n, the system is simulated to estimate the mean response time in steady state for one value of ρ, for example $\rho = 0.5$. This estimate obtained using simulations is denoted by $\widehat{\bar{T}}_5^{sim}(0.5)$. Using this estimate, Equation (4.2) can be solved to obtain an estimate of m_n represented by $\widehat{m}_n$. According to Conjecture 4.1, m_n is independent of ρ. Hence $\widehat{m}_n$ can now be used to estimate the mean response time for any other value of ρ in the interval $[0, 1)$. *Therefore, by simulating the system for one value of the traffic intensity* ρ, *it has become possible to obtain an estimate of the mean response time for infinitely many values of* ρ. This n-1 linear approximation algorithm for the symmetric n-dimensional fork-join queue with general service time is presented formally in Algorithm 1. The estimate of $\bar{T}_n$ obtained using the n-1 linear approximation algorithm for both exponential and general task service times is denoted by $\widehat{\bar{T}}_n$.

In the next section, the n-1 linear approximation algorithm to estimate the mean response time in steady state is demonstrated using numerical examples.

Algorithm 1 n-1 linear approximation algorithm for computation of $\widehat{\bar{T}}_n$

Input: Number of parallel tasks, n; Service rate, μ; Service time distribution $G(.)$

Output: Closed-form expression for the mean response time under any traffic intensity, $\rho \in [0, 1)$

1: Simulate the system for $\rho = 0.5$. Estimate the mean response time $\widehat{\bar{T}}_5^{sim}(0.5)$.

2: Compute the value of $\int_0^\infty nx\{G(x)\}^{n-1}dG(x)$

3: Solve this linear equation for $\widehat{m}_n$: $\widehat{\bar{T}}_5^{sim}(0.5) = \dfrac{0.5\widehat{m}_n}{\mu(1-0.5)} + \int_0^\infty nx\{G(x)\}^{n-1}dG(x)$

4: **return** $\widehat{\bar{T}}_n(\rho) = \dfrac{\rho\widehat{m}_n}{\mu(1-\rho)} + M_n$

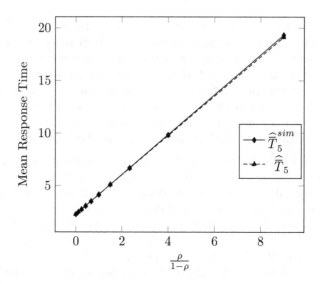

Figure 4.4: Plot of simulated and approximated (using Conjecture 4.2) mean response times vs. $\frac{\rho}{1-\rho}$ for $n = 5$, $\mu = 1$ and exponential service time distribution

4.2.3 Numerical Examples

Consider a symmetric n-dimensional fork-join queueing system with $n = 5$, $\mu = 1$ and exponential service time distribution. Conjecture 4.2 leads to the following approximation: $\widehat{\overline{T}}_5(\rho) = \frac{1.8735\rho}{1(1-\rho)} + \frac{2.2833}{1}$. In the plot in Figure 4.4 this approximation is compared against the mean response times obtained using simulations for values of ρ between 0.1 to 0.9 at increments of 0.1. The closeness of the two lines being compared demonstrates the remarkable accuracy of the approximation.

Now consider a system with $n = 5$, $\mu = 1$ and Erlang-2 service time distribution. Equation (4.2) results in the following linear relationship:

$$\overline{T}_5(\rho) = \frac{\rho m_5}{(1 - \rho)} + 1.904 \qquad (4.8)$$

The parameter m_5 in Equation 4.8 is unknown. The value of $E[T_5](\rho)$ for only one value of ρ is needed to estimate the unknown parameter, m_5. Therefore, the system is simulated for $\rho = 0.5$ and the value of $\widehat{\overline{T}}_5^{sim}(0.5)$ is estimated to be equal to 3.149. The

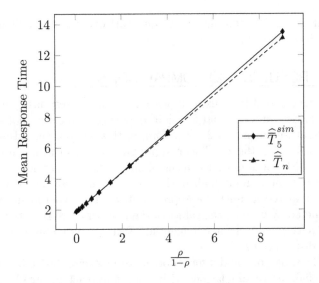

Figure 4.5: Plot of simulated and approximated (using Algorithm 1) mean response times vs. $\frac{\rho}{1-\rho}$ for $n = 5$, $\mu = 1$ and Erlang-2 service time distribution

simulation parameters used for obtaining the estimate of $\overline{T}_n$ are detailed in Section 4.3.

The value of $\widehat{\overline{T}}_5^{sim}$ (0.5) is substituted in Equation 4.8 and the following linear equation is solved for $\widehat{m}_n$:

$$3.149 = \widehat{m}_n + 1.904 \tag{4.9}$$

Equation 4.9 results in the following estimate of m_5: $\widehat{m}_5 = 1.245$. This estimate of m_5 can be now used for estimating the value of $\overline{T}_5(\rho)$ for any other value of the traffic intensity ρ as:

$$\widehat{\overline{T}}_n(\rho) = \frac{1.245\rho}{\mu(1 - \rho)} + 1.904 \tag{4.10}$$

Figure 4.5 shows the plot of the simulated average response times for ρ values of 0.1 to 0.9 at increments of 0.1 and the corresponding estimated value of the response time using Equation 4.10. The closeness of the approximation to the simulated values is demonstrated by the closeness between the two lines.

In the next section, an extensive comparison of the mean response time estimates obtained using: (i) simulations, (ii) the n-1 linear approximation algorithm, and (iii) existing approximation techniques in the literature is presented.

4.3   RESULTS AND COMPARISONS

Exact closed-form expressions for the mean response time of symmetric n-dimensional fork-join queueing systems do not exist in literature when $n > 2$. Therefore, in this section, the estimates obtained using the n-1 linear approximation algorithm are compared with those obtained using simulations. Furthermore, existing approximation methods in the literature are described in detail and compared against the approximation. The experiments were conducted by varying the values of the number of tasks n, the traffic intensity ρ, and for exponential, Erlang-2 and Pareto service time distributions.

The values of ρ and n for reporting results were chosen based on the reliability of simulations with respect to reaching steady state and having a reasonable variance of the response time realizations. For exponential and Erlang-2 service time distributions, results are reported for the traffic intensity ρ ranging from 0.1 to 0.9. For Pareto service time distribution, since the service time variability is higher, results are reported for ρ ranging from 0.1 to 0.8 when $n > 2$. When $n = 2$, results for $\rho = 0.9$ are also reported.

For exponential service time distribution, results are reported for $n = 3, 5, 10, 20, 30, 40$, and 50 (Tables (4.2)–(4.8)). For Erlang-2 service time distribution, results are reported for $n = 3, 5, 10, 15, 20$, and 30 (Tables (4.9)–(4.14)). Lastly, for Pareto service time distribution, results are reported for $n = 2, 3, 5, 7$ and 10 (Tables (4.15)–(4.19)). Without loss of generality, the mean service rate, μ, for all simulations and comparisons was assumed to take unit value, $i.e.$ $\mu = 1$. This defines the parameters for exponential and Erlang-2 service time distributions. The parameters used for the Pareto distribution were $\alpha = 2.291$ and $x_m = 0.563$.

In Tables (4.2)–(4.19), for different values of n, $\widehat{\overline{T}}_n^{sim}$ denotes the estimate of the mean response time in steady state obtained using simulations, $\widehat{\overline{T}}_n$ denotes the approximation using the n-1 linear approximation, $\widehat{\overline{T}}_n^{NT}$ denotes the approximation by Nelson and Tantawi [45], $\widehat{\overline{T}}_n^{TT}$ denotes the approximation by Thomasian and Tantawi [60], $\widehat{\overline{T}}_n^{VM}$ denotes the approximation by Varma and Makowski [70] and $\widehat{\overline{T}}_n^{KS}$ denotes the approximation by Ko and

Serfozo [33]. The approximations by these authors are described in Sections (4.3.2)–(4.3.5).

For each approximation technique, the error percentage, with the mean response time using simulations as the basis, is reported in the column next to the mean response time estimate obtained using the respective approximation. The error percentage calculation is the same for all the approximation methods compared. For example, for the n-1 linear approximation algorithm, the error percentage is calculated as:

$$\%Error = \frac{\left(\widehat{\overline{T}}_n - \widehat{\overline{T}}_n^{sim}\right) \times 100}{\widehat{\overline{T}}_n^{sim}} \tag{4.11}$$

The percentage error that is the least among the approximations compared is shown in bold font in Tables (4.2)–(4.19).

In the next section, a comparison of the mean response time estimate using the n-1 linear approximation with the mean response time estimated using simulations is discussed.

4.3.1 Comparison with Simulations

In this section, the simulation results reported are based on results from 10 simulation runs. In each simulation run, inter-arrival times of 30 million arriving jobs and the corresponding n service times for each job were generated. The response time of each job was calculated using Equation 4.1. Out of these, response times of the first 10 million jobs were not included in the calculation of the average to account for the warm-up period. Response times of the last 20 million jobs were used for the calculation of the average response time. 95% confidence intervals are reported for each average response time. These confidence intervals were generated using the average response times in the 10 simulation runs.

For the case of exponential service times (Tables (4.2)–(4.8)), the n-1 linear approximation estimate is obtained directly using Conjecture 4.2. The estimated mean response time $\widehat{\overline{T}}_n$ is remarkably close to the values observed in simulations with error percentages going only upto a maximum value of 3% even for the highest traffic intensity, $\rho = 0.9$ and highest number of tasks $n = 50$.

For Erlang-2 task service times (Tables (4.9)–(4.14)), the error percentages are within 4% for up to $\rho = 0.8$ and go up to 6% for $n = 20$ and $n = 30$ when $\rho = 0.9$.

When the task service times are according to a Pareto distribution (Tables (4.15)–(4.19)), the error percentages are less than 5% for up to $\rho = 0.8$ and less than 10% for $\rho = 0.9$ when $n = 7$ and $n = 10$. However, when the error percentages are higher than 5%, it is encouraging to note that $\widehat{\overline{T}}_n$ lies within the 95% confidence interval of $\widehat{\overline{T}}_n^{sim}$. The higher error percentages can be attributed to the higher variance of the Pareto distribution.

The increase in error percentages with increase in the traffic intensity ρ can be attributed to the higher warm-up period requirement to reach steady state for higher traffic intensities. Similarly, increase in the error percentage with increase in n can be attributed to the higher variance of the response time.

The time required for running the simulations ranged from a minimum of 16 minutes and 45 seconds to a maximum of 3 hours, 10 minutes and 20 seconds. Although these times are subject to change based on the specifications of the computing resource used, the simulation run times are unlikely to come close to the time required to estimate the mean response time based on the n-1 linear approximation algorithm. Using the n-1 linear approximation algorithm, $\widehat{\overline{T}}_n$ can be computed in less than a second for exponential service time distribution, while the other distributions require only one simulation to compute the estimate for other infinitely many values of $\rho \in [0,1)$. This advantage over simulations in terms of time and computing resources needed establishes the usefulness of the n-1 linear approximation algorithm in this comparison.

Approximations for the mean response time of the symmetric n-dimensional fork-join queueing system existing in literature are described and compared against in the next four sections.

4.3.2 Approximation by Nelson and Tantawi

Nelson and Tantawi [45] provided an approximation for the mean response time of a symmetric n-dimensional fork-join queueing system with exponential service time distribution. They observed that a lower bound to the expected response time is given by the maximum of n independent task service times, which for the exponential distribution is $\frac{H_n}{\mu}$. Using properties of associated random variables, they showed that an upper bound for the expected response time is given by the expected value of the maximum of the response times of n independent $M/M/1$ queues. This is given by $\frac{H_n}{\mu(1-\rho)}$. The authors used simulations to obtain a scaling approximation between these two bounds using the observation that both the lower and upper bounds increase with n at the same rate, H_n. The approximation obtained using this method is denoted by $\widehat{T}_n^{NT}$, and this is given by the following expression:

$$\widehat{T}_n^{NT} = \left[\frac{H_n}{H_2} + \frac{4}{11} \left(1 - \frac{H_n}{H_2} \right) \rho \right] \frac{12 - \rho}{8\mu(1-\rho)}. \qquad (4.12)$$

The estimated mean response time using Equation 4.12 and the corresponding error percentages are reported in Tables (4.2)–(4.8). Based on the results, the n-1 linear approximation technique scores over this method in the following ways:

1. The approximation using Equation 4.12 is valid only for exponential service time distribution. In comparison, the n-1 linear approximation algorithm is applicable to other distributions as well.

2. The n-1 linear approximation algorithm outperforms the approximation using Equation 4.12 in terms of error percentages in all the instances. In many cases, the difference in accuracy is significant. For example, in the case of $n = 50$ and $\rho = 0.9$, the percentage error of $\widehat{T}_{50}$ is -3.008% while the percentage error of $\widehat{T}_{50}^{NT}$ is 21.015% (Table 4.8).

3. The error percentages in all instances for the approximation using Equation 4.12 are positive. This points to the possibility that this method gives an upper bound rather than an approximation of the mean response time.

4.3.3 Approximation by Varma and Makowski

Varma and Makowski [70] considered a system with general inter-arrival and service times. They provided a light traffic approximation using the fact that the individual response times are independent when an arriving job encounters an empty system.

For systems with exponential inter-arrival times and general service times, they provided a conjecture for the heavy traffic limit using diffusion approximation. They extended this conjecture to the case of general inter-arrival and service times. This approximation is denoted by $\widehat{\overline{T}}_n^{VM}$. The heavy traffic conjecture is as follows:

$$\lim_{\rho \to 1} \mu(1 - \rho)\overline{T}_n(\rho) = \left[H_n + (4V_n - 3H_n - 1)\frac{\sigma_o^2}{\sigma_o^2 + \sigma^2} \right.$$

$$\left. + 2(1 + H_n - 2V_n)\left(\frac{\sigma_o^2}{\sigma_o^2 + \sigma^2}\right)^2 \right] \frac{\sigma_o^2 + \sigma^2}{2}\mu^2 \quad (4.13)$$

where σ_o^2 and σ^2 are the variances of the inter-arrival time and service time distributions respectively and $V_n = \sum_{r=1}^{n} \binom{n}{r}(-1)^{r-1} \sum_{m=1}^{r} \binom{r}{m}\frac{(m-1)!}{r^{m+1}}$.

Using the light traffic approximation and the heavy traffic conjecture, the authors obtained an interpolation approximation for the mean response time. In the experiments conducted, this approximation is compared against the n-1 linear approximation for exponential, Erlang-2 and Pareto service time distributions. For exponential service time distribution, $\widehat{\overline{T}}_n^{VM}$ is given by the following:

$$\widehat{\overline{T}}_n^{VM} = \left[H_n + (V_n - H_n)\rho \right]\frac{1}{\mu(1 - \rho)}. \quad (4.14)$$

For Erlang-2 service time distribution with mean equal to "1", $\widehat{\overline{T}}_n^{VM}$ is as follows:

$$\widehat{\overline{T}}_n^{VM} = \left[F_n + \left(\frac{1}{6} - \frac{H_n}{12} + \frac{2}{3}V_n - F_n\right)\rho \right]\frac{1}{\mu(1 - \rho)}, \quad n = 2, 3, \ldots \quad (4.15)$$

where $F_n = \sum_{r=1}^{n} \binom{n}{r}(-1)^{r-1} \sum_{m=0}^{r} \binom{r}{m}\frac{(m)!}{2r^{m+1}}$.

For Pareto service time distribution, $\widehat{\overline{T}}_n^{VM}$ is given by the following expression:

$$\widehat{\overline{T}}_n^{VM} = \frac{M_n + g_1\mu\rho}{\mu(1 - \rho)} \quad (4.16)$$

where $g_1 = \left[H_n + (4V_n - 3H_n - 1)\frac{\sigma_o^2}{\sigma_o^2 + \sigma^2} + 2(1 + H_n - 2V_n)\left(\frac{\sigma_o^2}{\sigma_o^2 + \sigma^2}\right)^2 \right] \frac{\sigma_o^2 + \sigma^2}{2}\mu^2 - M_n$.

Based on results in Tables (4.2)–(4.19), the n-1 linear approximation algorithm compares against this technique as follows:

1. For exponential and Erlang-2 service time distributions, the n-1 linear approximation performs better in terms of the percentage error for higher values of n and ρ. For example, when $n = 50$, $\rho = 0.9$ and service times are exponentially distributed, the error percentage of $\widehat{T}_{50}$ is -3.008% while the error percentage of $\widehat{T}_{50}^{VM}$ is 9.556 (Table 4.8). For the cases where $\widehat{T}_n^{VM}$ performs better, which is generally observed for lower values of ρ, the error percentages of both $\widehat{T}_n^{VM}$ and $\widehat{T}_n$ are less than 1% and the performance is comparable.

2. This technique performs well only when the service time distributions are exponential-like. When the service times are distributed according to a Pareto distribution, the utility of the approximation using Equation 4.16 is questionable. The error percentages go up to -44.752% when $n = 10$ and $\rho = 0.8$ (Table 4.19). In comparison, the error percentage of $\widehat{T}_n$ is 9.643%.

3. The accuracy of the approximation $\widehat{T}_n^{VM}$ for low values of ρ can be explained by the fact that this is an interpolation approximation and the exact mean response time is known when $\rho = 0^+$.

4. Conjectures 4.1 and 4.2 contradict the heavy traffic conjecture of Varma and Makowski [70]. The lower error percentages of $\widehat{T}_n$ lends more confidence to Conjectures 4.1 and 4.2.

4.3.4 Approximation by Ko and Serfozo

Ko and Serfozo [33] proposed an approximation for the expected response time in a fork-join queueing system with exponential inter-arrival and service times, where each node acts as

Table 4.2: Comparison with simulation results and approximations by Nelson and Tantawi [45], Varma and Makowski [70], and Ko and Serfozo [33] for $n = 3$ and exponential service time distribution

ρ	$\widehat{T}_3^{sim}$	$\widehat{T}_3$	% Error	$\widehat{T}_3^{NT}$	% Error	$\widehat{T}_3^{VM}$	% Error	$\widehat{T}_3^{KS}$	% Error
0.1	2.01±0.000	2.01	**-0.002**	2.01	0.180	2.011	0.009	2.21	10.131
0.2	2.23±0.000	2.23	-0.020	2.22	0.373	2.232	**-0.005**	2.49	11.586
0.3	2.51±0.000	2.52	-0.050	2.50	0.573	2.517	**-0.034**	2.85	13.065
0.4	2.89±0.001	2.90	-0.0921	2.88	0.780	2.896	**-0.072**	3.32	14.573
0.5	3.43±0.001	3.43	**-0.141**	3.40	1.014	3.428	-0.146	3.99	16.114
0.6	4.23±0.001	4.22	-0.207	4.18	1.218	4.225	**-0.201**	4.98	17.677
0.7	5.57±0.003	5.55	-0.307	5.49	1.459	5.553	**-0.298**	6.64	19.245
0.8	8.24±0.007	8.21	-0.453	8.10	1.694	8.210	**-0.397**	9.96	20.803
0.9	16.27±0.025	16.18	**-0.582**	15.95	2.122	16.181	-0.703	19.93	22.429

a queueing station with s servers, $s \geq 1$. They used elementary probability laws and properties of associated random variables to derive this approximation. This approximation for the number of servers $s = 1$ is denoted by $\widehat{T}_n^{KS}$. When the number of tasks $n = 2$, this approximation is the same as the exact result by Nelson and Tantawi [45]. For $n \geq 3$, $\widehat{T}_n^{KS}$ is given by:

$$\widehat{T}_n^{KS} = \left[\left(\frac{3}{2} - \frac{\rho}{4} \right) + \left(\frac{5}{4} - \frac{\rho}{8} \right) \left(\frac{5}{4} - \frac{\rho}{4} \right) \left(H_n - \frac{3}{2} \right) \right] \frac{1}{\mu(1 - \rho)}$$

$$(4.17)$$

The n-1 linear approximation algorithm scores over this method for the same reasons as enumerated in Section 4.3.2. However, in this case, the error percentages of $\widehat{T}_n^{KS}$ are exorbitantly high, even for low values of ρ and n. When $n = 3$ and $\rho = 0.1$, for exponential service time distribution (Table 4.2), the error percentage is 10.131% which is the lowest observed. This goes up to 79.818% when $n = 50$ and $\rho = 0.9$ making the utility of this approximation questionable.

4.3.5 Approximation by Thomasian and Tantawi

Thomasian and Tantawi [60] extended the work of Nelson and Tantawi [45] to propose an approximation for the expected response time of symmetric fork-join queues with exponential inter-arrival times and general service times. This approximation is denoted by $\widehat{T}_n^{TT}$ and is given by:

Table 4.3: Comparison with simulation results and approximations by Nelson and Tantawi [45], Varma and Makowski [70], and Ko and Serfozo [33] for $n = 5$ and exponential service time distribution

ρ	$\widehat{T}_5^{sim}$	$\widehat{T}_5$	% Error	$\widehat{T}_5^{NT}$	% Error	$\widehat{T}_5^{VM}$	% Error	$\widehat{T}_5^{KS}$	% Error
0.1	2.49±0.000	2.49	0.047	2.50	0.503	2.49	-0.001	2.97	19.365
0.2	2.75±0.000	2.75	0.051	2.78	0.989	2.75	-0.044	3.34	21.607
0.3	3.08±0.000	3.09	0.024	3.13	1.471	3.08	-0.121	3.82	23.897
0.4	3.53±0.001	3.53	-0.045	3.60	1.940	3.52	-0.243	4.46	26.223
0.5	4.16±0.001	4.15	-0.164	4.26	2.388	4.14	-0.418	5.35	28.578
0.6	5.11±0.002	5.09	-0.343	5.25	2.805	5.07	-0.653	6.69	30.949
0.7	6.69±0.003	6.65	-0.568	6.90	3.207	6.62	-0.935	8.92	33.356
0.8	9.85±0.009	9.77	-0.845	10.20	3.590	9.72	-1.273	13.37	35.793
0.9	19.35±0.033	19.12	-1.176	20.11	3.950	19.03	-1.666	26.75	38.260

$$\widehat{T}_n^{TT} = \bar{T}_1 + \frac{\int_0^\infty nx\{G(x)\}^{n-1}dG(x) - \frac{1}{\mu}}{\sigma_G}\alpha_n(\rho) \qquad (4.18)$$

where, $\bar{T}_1$, μ, and $G(x)$ are as defined in Sections 4.1 and 4.2 respectively, σ_G is the standard deviation of $G(x)$, and $\alpha_n(\rho)$ is a possibly non-linear function of the traffic intensity ρ and depends on $G(x)$. Estimation of $\alpha_n(\rho)$ requires multiple simulations to obtain enough data points to be able to fit a function on the values of $\alpha_n(\rho)$.

The advantages of the n-1 linear approximation over the approximation by Thomasian and Tantawi [60] are as follows:

1. The computation of $\widehat{T}_n^{TT}$ requires multiple simulations to fit a curve on $\alpha_n(\rho)$. However, the n-1 linear approximation requires no simulations for the case of exponential service times and exactly one simulation for other service time distributions. Since simulations consume significant time and computing resources, therefore the n-1 linear approximation is more resource-efficient.

2. The n-1 linear approximation performs significantly better in terms of the error percentages observed in Tables (4.9)–(4.14). For example, when $n = 15$ and $\rho = 0.9$ (Table 4.12), the error percentage of $\widehat{T}_{15}^{TT}$ is -35.127 while that of $\widehat{T}_{15}$ is 4.909.

The discussion presented above establishes the superior performance of the n-1 linear approximation when compared to approximations existing in the literature.

Table 4.4: Comparison with simulation results and approximations by Nelson and Tantawi [45], Varma and Makowski [70], and Ko and Serfozo [33] for $n = 10$ and exponential service time distribution

ρ	$\widehat{T}_{10}^{sim}$	$\widehat{T}_{10}$	% Error	$\widehat{T}_{10}^{NT}$	% Error	$\widehat{T}_{10}^{VM}$	% Error	$\widehat{T}_{10}^{KS}$	% Error
0.1	3.17±0.000	3.18	0.143	3.21	1.259	3.17	**-0.031**	4.06	27.906
0.2	3.48±0.000	3.49	0.226	3.57	2.523	3.48	**-0.127**	4.57	31.145
0.3	3.88±0.001	3.89	**0.236**	4.03	3.780	3.87	-0.305	5.22	34.448
0.4	4.42±0.001	4.43	**0.161**	4.64	5.019	4.39	-0.579	6.09	37.804
0.5	5.18±0.002	5.17	**-0.022**	5.50	6.219	5.13	-0.970	7.31	41.185
0.6	6.32±0.002	6.30	**-0.323**	6.78	7.370	6.22	-1.488	9.13	44.577
0.7	8.23±0.005	8.17	**-0.751**	8.93	8.463	8.05	-2.143	12.18	47.970
0.8	12.07±0.014	11.91	**-1.348**	13.22	9.453	11.72	-2.975	18.27	51.298
0.9	23.64±0.063	23.14	**-2.134**	26.08	10.314	22.70	-4.003	36.54	54.524

Table 4.5: Comparison with simulation results and approximations by Nelson and Tantawi [45], Varma and Makowski [70], and Ko and Serfozo [33] for $n = 20$ and exponential service time distribution

ρ	$\widehat{T}_{20}^{sim}$	$\widehat{T}_{20}$	% Error	$\widehat{T}_{20}^{NT}$	% Error	$\widehat{T}_{20}^{VM}$	% Error	$\widehat{T}_{20}^{KS}$	% Error
0.1	3.88±0.000	3.89	0.289	3.95	1.885	3.88	**-0.047**	5.19	33.743
0.2	4.23±0.001	4.25	0.490	4.39	3.786	4.22	**-0.203**	5.83	37.858
0.3	4.69±0.001	4.72	0.578	4.96	5.683	4.67	**-0.491**	6.67	42.059
0.4	5.32±0.001	5.34	**0.533**	5.72	7.555	5.27	-0.938	7.78	46.321
0.5	6.20±0.002	6.22	**0.326**	6.78	9.378	6.10	-1.566	9.34	50.608
0.6	7.53±0.003	7.53	**-0.072**	8.37	11.118	7.35	-2.405	11.67	54.878
0.7	9.78±0.005	9.71	**-0.698**	11.03	12.734	9.44	-3.494	15.56	59.067
0.8	14.31±0.010	14.08	**-1.604**	16.34	14.163	13.61	-4.881	23.34	63.084
0.9	27.96±0.040	27.18	**-2.787**	32.27	15.400	26.13	-6.560	46.68	66.915

Table 4.6: Comparison with simulation results and approximations by Nelson and Tantawi [45], Varma and Makowski [70], and Ko and Serfozo [33] for $n = 30$ and exponential service time distribution

ρ	$\widehat{T}_{30}^{sim}$	$\widehat{T}_{30}$	% Error	$\widehat{T}_{30}^{NT}$	% Error	$\widehat{T}_{30}^{VM}$	% Error	$\widehat{T}_{30}^{KS}$	% Error
0.1	4.29±0.000	4.31	0.381	4.39	2.197	4.29	**-0.058**	5.86	36.354
0.2	4.67±0.001	4.71	0.659	4.88	4.418	4.66	**-0.249**	6.59	40.926
0.3	5.17±0.001	5.21	0.805	5.51	6.639	5.14	**-0.600**	7.53	45.598
0.4	5.84±0.001	5.89	**0.791**	6.36	8.834	5.78	-1.144	8.78	50.337
0.5	6.79±0.002	6.84	**0.586**	7.54	10.973	6.67	-1.911	10.54	55.101
0.6	8.24±0.003	8.26	**0.153**	9.32	13.016	8.00	-2.935	13.17	59.834
0.7	10.68±0.005	10.62	**-0.544**	12.27	14.921	10.23	-4.251	17.57	64.475
0.8	15.59±0.013	15.36	**-1.535**	18.19	16.646	14.68	-5.887	26.35	68.959
0.9	30.39±0.043	29.56	**-2.755**	35.94	18.266	28.03	-7.779	52.70	73.387

Table 4.7: Comparison with simulation results and approximations by Nelson and Tantawi [45], Varma and Makowski [70], and Ko and Serfozo [33] for $n = 40$ and exponential service time distribution

ρ	$\widehat{T}^{sim}_{40}$	$\widehat{T}_{40}$	% Error	$\widehat{T}^{NT}_{40}$	% Error	$\widehat{T}^{VM}_{40}$	% Error	$\widehat{T}^{KS}_{40}$	% Error
0.1	4.59±0.000	4.61	0.449	4.70	2.400	4.59	**-0.065**	6.33	37.948
0.2	4.99±0.000	5.03	0.782	5.23	4.827	4.97	**-0.279**	7.12	42.820
0.3	5.51±0.001	5.56	0.971	5.91	7.257	5.47	**-0.678**	8.14	47.801
0.4	6.21±0.001	6.28	**0.978**	6.82	9.657	6.13	-1.295	9.50	52.847
0.5	7.22±0.001	7.27	**0.771**	8.08	11.994	7.06	-2.165	11.40	57.914
0.6	8.74±0.002	8.77	**0.312**	9.99	14.225	8.45	-3.324	14.25	62.942
0.7	11.32±0.004	11.27	**-0.446**	13.16	16.296	10.77	-4.816	19.00	67.847
0.8	16.51±0.011	16.26	**-1.510**	19.51	18.190	15.41	-6.646	28.50	72.602
0.9	32.19±0.069	31.24	**-2.959**	38.57	19.802	29.33	-8.886	57.00	77.047

Table 4.8: Comparison with simulation results and approximations by Nelson and Tantawi [45], Varma and Makowski [70], and Ko and Serfozo [33] for $n = 50$ and exponential service time distribution

ρ	$\widehat{T}^{sim}_{50}$	$\widehat{T}_{50}$	% Error	$\widehat{T}^{NT}_{50}$	% Error	$\widehat{T}^{VM}_{50}$	% Error	$\widehat{T}^{KS}_{50}$	% Error
0.1	4.82±0.000	4.84	0.500	4.94	2.545	4.82	**-0.062**	6.70	39.059
0.2	5.23±0.000	5.28	0.877	5.50	5.123	5.22	**-0.289**	7.54	44.152
0.3	5.77±0.001	5.83	1.100	6.22	7.707	5.73	**-0.713**	8.62	49.362
0.4	6.50±0.001	6.58	**1.130**	7.17	10.260	6.41	-1.373	10.06	54.640
0.5	7.55±0.002	7.62	**0.921**	8.51	12.738	7.37	-2.314	12.07	59.926
0.6	9.13±0.003	9.17	**0.440**	10.51	15.100	8.81	-3.570	15.09	65.160
0.7	11.81±0.006	11.77	**-0.351**	13.86	17.302	11.20	-5.174	20.11	70.276
0.8	17.22±0.016	16.96	**-1.471**	20.54	19.314	15.99	-7.142	30.17	75.225
0.9	33.56±0.086	32.55	**-3.008**	40.61	21.015	30.35	-9.556	60.34	79.818

Table 4.9: Comparison with simulation results and approximations by Thomasian and Tantawi [45] and Varma and Makowski [70] for $n = 3$ and Erlang-2 service time distribution

ρ	$\widehat{T}^{sim}_{3}$	$\widehat{T}_{3}$	% Error	$\widehat{T}^{TT}_{3}$	% Error	$\widehat{T}^{VM}_{3}$	% Error
0.1	1.73±0.000	1.73	0.100	1.93	11.658	1.73	**0.003**
0.2	1.88±0.000	1.88	0.163	2.03	7.966	1.88	**-0.037**
0.3	2.07±0.000	2.07	0.176	2.17	4.566	2.07	**-0.130**
0.4	2.33±0.000	2.33	**0.127**	2.37	1.463	2.32	-0.297
0.6	3.25±0.001	3.24	**-0.240**	3.13	-3.854	3.22	-0.920
0.7	4.18±0.003	4.15	**-0.622**	3.93	-6.080	4.12	-1.445
0.8	6.04±0.007	5.97	**-1.208**	5.56	-8.051	5.91	-2.184
0.9	11.67±0.031	11.43	**-2.041**	10.53	-9.778	11.30	-3.179

Table 4.10: Comparison with simulation results and approximations by Thomasian and Tantawi [45] and Varma and Makowski [70] for $n = 5$ and Erlang-2 service time distribution

ρ	$\widehat{\overline{T}}_5^{sim}$	$\widehat{\overline{T}}_5$	% Error	$\widehat{\overline{T}}_5^{TT}$	% Error	$\widehat{\overline{T}}_5^{VM}$	% Error
0.1	2.04±0.000	2.04	0.154	2.33	14.013	2.04	**0.002**
0.2	2.21±0.000	2.22	0.251	2.40	8.675	2.21	**-0.064**
0.3	2.43±0.000	2.44	0.274	2.52	3.655	2.43	**-0.220**
0.4	2.73±0.000	2.73	**0.200**	2.70	-1.051	2.72	-0.486
0.6	3.78±0.001	3.77	**-0.371**	3.42	-9.552	3.73	-1.483
0.7	4.85±0.003	4.81	**-0.954**	4.21	-13.372	4.74	-2.303
0.8	7.01±0.006	6.88	**-1.809**	5.82	-16.929	6.77	-3.411
0.9	13.51±0.028	13.11	**-2.992**	10.78	-20.238	12.85	-4.863

Table 4.11: Comparison with simulation results and approximations by Thomasian and Tantawi [45] and Varma and Makowski [70] for $n = 10$ and Erlang-2 service time distribution

ρ	$\widehat{\overline{T}}_{10}^{sim}$	$\widehat{\overline{T}}_{10}$	% Error	$\widehat{\overline{T}}_{10}^{TT}$	% Error	$\widehat{\overline{T}}_{10}^{VM}$	% Error
0.1	2.47±0.000	2.47	0.228	2.87	16.245	2.47	**-0.030**
0.2	2.66±0.000	2.67	0.369	2.91	9.325	2.66	**-0.171**
0.3	2.92±0.000	2.93	**0.402**	3.00	2.741	2.91	-0.440
0.4	3.26±0.001	3.27	**0.293**	3.15	-3.517	3.24	-0.879
0.6	4.50±0.001	4.48	**-0.539**	3.82	-15.119	4.39	-2.448
0.7	5.76±0.002	5.68	**-1.390**	4.58	-20.509	5.55	-3.711
0.8	8.31±0.006	8.09	**-2.631**	6.18	-25.659	7.86	-5.391
0.9	16.00±0.037	15.31	**-4.330**	11.11	-30.584	14.79	-7.553

Table 4.12: Comparison with simulation results and approximations by Thomasian and Tantawi [45] and Varma and Makowski [70] for $n = 15$ and Erlang-2 service time distribution

ρ	$\widehat{\overline{T}}_{15}^{sim}$	$\widehat{\overline{T}}_{15}$	% Error	$\widehat{\overline{T}}_{15}^{TT}$	% Error	$\widehat{\overline{T}}_{15}^{VM}$	% Error
0.1	2.71±0.000	2.72	0.266	3.18	17.210	2.71	**-0.062**
0.2	2.93±0.000	2.94	0.436	3.21	9.598	2.92	**-0.250**
0.3	3.20±0.000	3.22	**0.477**	3.28	2.325	3.18	-0.602
0.4	3.57±0.000	3.59	**0.349**	3.41	-4.622	3.53	-1.152
0.6	4.92±0.001	4.88	**-0.626**	4.05	-17.609	4.76	-3.085
0.7	6.28±0.002	6.18	**-1.598**	4.79	-23.685	5.99	-4.589
0.8	9.05±0.007	8.78	**-2.979**	6.38	-29.503	8.46	-6.541
0.9	17.42±0.026	16.56	**-4.909**	11.30	-35.127	15.84	-9.072

Table 4.13: Comparison with simulation results and approximations by Thomasian and Tantawi [45] and Varma and Makowski [70] for $n = 20$ and Erlang-2 service time distribution

ρ	$\widehat{T}_{20}^{sim}$	$\widehat{T}_{20}$	% Error	$\widehat{T}_{20}^{TT}$	% Error	$\widehat{T}_{20}^{VM}$	% Error
0.1	2.89±0.000	2.90	0.289	3.41	17.791	2.89	**-0.091**
0.2	3.11±0.000	3.13	0.480	3.42	9.766	3.10	**-0.318**
0.3	3.40±0.000	3.42	**0.529**	3.47	2.080	3.38	-0.723
0.4	3.79±0.001	3.81	**0.386**	3.59	-5.286	3.74	-1.362
0.6	5.21±0.002	5.17	**-0.697**	4.21	-19.125	5.02	-3.561
0.7	6.65±0.004	6.53	**-1.778**	4.95	-25.642	6.30	-5.266
0.8	9.58±0.010	9.26	**-3.339**	6.52	-31.928	8.86	-7.491
0.9	18.44±0.043	17.44	**-5.431**	11.43	-37.981	16.54	-10.285

Table 4.14: Comparison with simulation results and approximations by Thomasian and Tantawi [45] and Varma and Makowski [70] for $n = 30$ and Erlang-2 service time distribution

ρ	$\widehat{T}_{30}^{sim}$	$\widehat{T}_{30}$	% Error	$\widehat{T}_{30}^{TT}$	% Error	$\widehat{T}_{30}^{VM}$	% Error
0.1	3.14±0.000	3.15	0.328	3.72	18.522	3.13	**-0.126**
0.2	3.37±0.000	3.39	0.539	3.71	10.002	3.36	**-0.410**
0.3	3.68±0.000	3.70	**0.592**	3.74	1.819	3.64	-0.900
0.4	4.10±0.001	4.11	**0.432**	3.85	-6.049	4.03	-1.651
0.6	5.61±0.003	5.57	**-0.776**	4.44	-20.916	5.37	-4.198
0.7	7.16±0.006	7.02	**-1.984**	5.16	-27.964	6.72	-6.155
0.8	10.30±0.014	9.92	**-3.714**	6.72	-34.785	9.41	-8.682
0.9	19.83±0.051	18.63	**-6.048**	11.62	-41.392	17.48	-11.856

Table 4.15: Comparison with simulation results and approximation by Varma and Makowski [70] for $n = 2$ and Pareto service time distribution

ρ	$\widehat{T}_{2}^{sim}$	$\widehat{T}_{2}$	% Error	$\widehat{T}_{2}^{VM}$	% Error
0.1	1.50±0.011	1.49	**-0.134**	1.48	-1.297
0.2	1.77±0.026	1.76	**-0.221**	1.72	-2.437
0.3	2.11±0.045	2.11	**-0.250**	2.04	-3.422
0.4	2.57±0.069	2.57	**-0.159**	2.46	-4.214
0.6	4.17±0.156	4.18	**0.247**	3.94	-5.383
0.7	5.75±0.244	5.79	**0.653**	5.42	-5.693
0.8	8.87±0.424	9.00	**1.519**	8.38	-5.530
0.9	18.10±0.982	18.6609	**3.111**	17.25	-4.663

Table 4.16: Comparison with simulation results and approximation by Varma and Makowski [70] for $n = 3$ and Pareto service time distribution

ρ	$\widehat{T}_3^{sim}$	$\widehat{T}_3$	% Error	$\widehat{T}_3^{VM}$	% Error
0.1	1.82±0.052	1.81	**-0.386**	1.73	-4.860
0.2	2.21±0.114	2.20	**-0.543**	2.02	-8.808
0.3	2.72±0.195	2.71	**-0.567**	2.39	-12.097
0.4	3.39±0.304	3.38	**-0.401**	2.89	-14.795
0.6	5.69±0.681	5.73	**0.683**	4.63	-18.620
0.7	7.96±1.076	8.08	**1.510**	6.37	-19.956
0.8	12.42±1.881	12.78	**2.854**	9.85	-20.716

Table 4.17: Comparison with simulation results and approximation by Varma and Makowski [70] for $n = 5$ and Pareto service time distribution

ρ	$\widehat{T}_5^{sim}$	$\widehat{T}_5$	% Error	$\widehat{T}_5^{VM}$	% Error
0.1	2.30±0.051	2.28	**-0.648**	2.12	-7.936
0.2	2.86±0.112	2.84	**-0.935**	2.46	-14.096
0.3	3.58±0.192	3.55	**-0.980**	2.90	-19.010
0.4	4.53±0.297	4.50	**-0.702**	3.49	-22.894
0.6	7.73±0.658	7.81	**1.034**	5.55	-28.197
0.7	10.85±1.038	11.13	**2.602**	7.62	-29.810
0.8	16.87±1.792	17.77	**5.307**	11.74	-30.422

Table 4.18: Comparison with simulation results and approximation by Varma and Makowski [70] for $n = 7$ and Pareto service time distribution

ρ	$\widehat{T}_7^{sim}$	$\widehat{T}_7$	% Error	$\widehat{T}_7^{VM}$	% Error
0.1	2.66±0.033	2.63	**-0.918**	2.42	-8.977
0.2	3.33±0.074	3.28	**-1.371**	2.80	-15.868
0.3	4.17±0.129	4.11	**-1.405**	3.28	-21.229
0.4	5.27±0.199	5.22	**-0.953**	3.93	-25.359
0.6	8.95±0.439	9.09	**1.554**	6.20	-30.753
0.7	12.46±0.679	12.97	**4.049**	8.47	-32.055
0.8	19.16±1.176	20.72	**8.105**	13.00	-32.145

Table 4.19: Comparison with simulation results and approximation by Varma and Makowski [70] for $n = 10$ and Pareto service time distribution

ρ	$\widehat{T}_{10}^{sim}$	$\widehat{T}_{10}$	% Error	$\widehat{T}_{10}^{VM}$	% Error
0.1	3.23±0.075	3.18	-1.383	2.79	-13.546
0.2	4.17±0.162	4.09	-1.887	3.21	-23.091
0.3	5.35±0.274	5.25	-1.798	3.74	-30.130
0.4	6.88±0.425	6.80	-1.194	4.44	-35.431
0.6	12.00±0.957	12.23	1.928	6.92	-42.278
0.7	16.83±1.491	17.65	4.891	9.40	-44.124
0.8	26.00±2.577	28.50	9.643	14.36	-44.752

4.4 CONCLUSIONS

In this chapter, an analysis of the symmetric n-dimensional fork-join queueing system is presented. Even though this queueing system has been studied in the literature for more than thirty years, its performance analysis has remained a very difficult problem. This is due to dependence between random variables that arises as a result of synchronized arrivals to the parallel queues. A conjecture on the relationship between the mean response time of the system and the mean response time of a single queue in the system is presented. The conjecture, which is strongly supported by simulations, is used to formulate the n-1 linear approximation algorithm. This algorithm estimates the mean response time of the symmetric n-dimensional fork-join queueing system with remarkable accuracy. The n-1 linear approximation is compared against other approximations proposed in the literature and compelling arguments and results in favor of the former are presented.

The utility of the analysis presented in this chapter lies in making informed decisions on system design. The fork-join queueing system is stable when the arrival rate of jobs is less than the service rate. However, the response time increases as a continuous function of the traffic intensity and converges to infinity as the traffic intensity approaches the value of one. Consequently, even when the system is stable, the response times can become inordinately large as the arrival rate increases and gets closer to the value of the service rate. However, given a target mean response time of the system, the algorithm formulated in this chapter can be used to estimate the maximum allowable arrival rate to the system by simply solving a linear equation. In the next chapter, the ideas presented in this chapter are extended to more complex forms of fork-join queueing networks.

CHAPTER 5

Relaxed Fork-Join Queueing Networks

In this chapter, three relaxations of the symmetric n-dimensional fork-join queueing system are analyzed: (i) symmetric tandem fork-join queues, (ii) heterogeneous fork-join queues, and (iii) (n, k) fork-join queues. For each queueing system, a linear relationship between the mean response time of the system and the mean response time of a single task queue is conjectured. The conjectures are supported by extensive simulations. Approximations based on the conjectures are proposed to estimate the mean response time of symmetric tandem fork-join queues and heterogeneous fork-join queues in steady state.

T HE SYMMETRIC n-dimensional fork-join queue is the most analytically tractable form of fork-join queueing networks. This system is useful for modeling some applications, such as wireless sensor networks where all the sensors are identical. However, it can be restrictive for other applications. For example, consider a manufacturing facility where components of products can be manufactured in parallel production lines. The production line of each component of the product might consist of several queueing stations, each providing a different service, for example, filing or welding. Such a system would require more than one single level of tasks in the fork-join queueing network. Furthermore, since each component might be different, the parallel production lines might not be stochastically identical, which would require allowing for

DOI: 10.1201/9781003150077-5

non-symmetrical or heterogeneous tasks. In other applications such as network coding, there might be redundancy between parallel tasks and completion of all the n tasks might not be required for job completion.

Keeping such applications in mind, in this chapter, the following three aspects of the symmetric n-dimensional fork-join queueing system are relaxed and analyzed individually: (i) the queueing network structure, (ii) symmetry of tasks, and (iii) the number of tasks that need to be complete before the job exits the system.

In the fork-join queueing system analyzed in Section 5.1, the restriction on the network structure is relaxed to analyze symmetric tandem fork-join queueing networks. In Section 5.2, the assumption of symmetric tasks is relaxed to analyze heterogeneous fork-join queueing systems. Finally, in Section 5.3, the system design of (n, k) fork-join queues only requires k of the n tasks to finish service for the job to be complete.

5.1 SYMMETRIC TANDEM FORK-JOIN QUEUEING NETWORK

In a symmetric tandem fork-join queueing network, jobs arrive according to a Poisson process with rate λ. On arrival, each job splits into n tasks. Each task consists of l sub-tasks in series, each of which is served at its own queueing station, which consists of a single server operating under FCFS service discipline. Completion of task i corresponds to service completion at the last of the l sub-task queues constituting task i, $1 \leq i \leq n$. After completion of all the n tasks, the job is complete and it departs from the system. The service time at each station is exponentially distributed with mean $1/\mu$. These service times are independent across sub-tasks as well as jobs. This fork-join network is shown in Figure 5.1. The symmetric tandem fork-join queueing network falls into a wider class of queueing networks known as acyclic fork-join queueing networks. Similar to the symmetric n-dimensional fork-join queue, Baccelli et al [7] show that this system is also stable *iff* $\lambda < \mu$. The authors derive bounds for the mean response time, which are applicable to symmetric tandem fork-join queues. Heavy traffic approximations proposed by Nguyen et al [47] are also applicable to this system. However, no approximations are available in literature for general traffic intensities for this system.

In the next section, the n-1 linear approximation algorithm from Chapter 4 is extended to estimate the mean response time of the symmetric tandem fork-join queueing network in steady state.

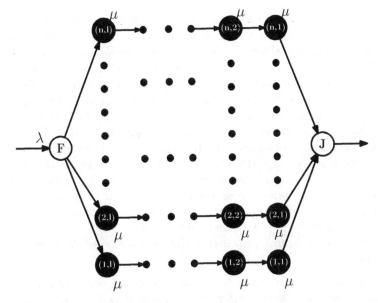

Figure 5.1: A symmetric tandem fork-join queueing network

5.1.1 Response Time Estimation

The following conjecture forms the basis for estimation of the mean response time of a symmetric tandem fork-join queueing network in steady state.

Conjecture 5.1 *The mean response time of a symmetric tandem fork-join queueing network in steady state, with exponential inter-arrival and sub-task service time distributions, is linear with respect to $\frac{\rho}{1-\rho}$, i.e. the slope and intercept are independent of the arrival rate, λ and the service rate, μ.*

$$E[T_{(n,l)}] = \frac{m_{(n,l)}\rho}{\mu(1-\rho)} + \frac{M_{(n,l)}}{\mu} \tag{5.1}$$

where, $m_{(n,l)}$ is a parameter independent of λ and $M_{(n,l)}$ is the mean of the maximum of n iid $Erlang(l,1)$ random variables.

Intuition behind Conjecture 5.1 Let R_l denote the number of sub-task queueing stations remaining for the final task when the last among the first $n-1$ tasks is complete. The expected value of R_l is the difference between l and the expected number of sub-task stations where service is complete in the slowest task queue at the first instant of time when the first $n-1$ tasks are complete. The length of the time interval between the arrival of the job and when the value of R_l is observed depends on the time needed to complete the first $n-1$ tasks of the job. Since the arrivals are synchronized and all the sub-tasks have identical service time distribution, if the arrival intensity changes, it affects all the sub-tasks uniformly. It is conjectured that the time interval after which R_l is observed changes with the change in arrival rate in such a way that R_l does not depend on the arrival rate. In fact Lemma 5.1 proves that this is true when $n=2$. If this holds true, then the time required at the remaining R_l sub-task queueing stations has a similar relationship to the traffic intensity ρ as the synchronization time of the last task in a symmetric n-dimensional fork-join queue *i.e.* linearity with respect to $\frac{\rho}{1-\rho}$. This leads to Conjecture 5.1.

The next lemma proves that the expectation of R_l is indeed independent of λ and μ when $n=2$.

Lemma 5.1 *In a symmetric tandem fork-join queueing network in steady state, with number of tasks, $n=2$ and number of sub-tasks within each task, $l \geq 1$, the expected number of sub-tasks of the second task remaining unfinished at the first instant of time when the first task is complete, R_l, is independent of the arrival intensity λ and is given by:*

$$E[R_l] = \sum_{s=1}^{l} \frac{s}{2^{2l-s}} \binom{2l-s-1}{l-s} = M_{(2,l)} - l - 1 \qquad (5.2)$$

Proof 5.1 *Let $T^{(1)}$ denote the time required to complete the first task in steady state. Each sub-task in the second task requires a random amount of time that is exponentially distributed with rate $\mu(1-\rho)$. Conditioned on knowing $T^{(1)}$, the number of completed sub-tasks of the second task is a Poisson random variable. The expectation is given by the following:*

$$E[R_l] = E\Big[E[R_l|T^{(1)}]\Big] = E\left[\sum_{s=1}^{l} s \cdot \frac{e^{-\mu(1-\rho)T^{(1)}}\{\mu(1-\rho)T^{(1)}\}^{l-s}}{(l-s)!}\right]$$

$$(5.3)$$

Since the distribution of $T^{(1)}$ is known to be Erlang $(l, \mu(1-\rho))$, the following holds true:

$$E\left[\frac{e^{-\mu(1-\rho)T^{(1)}}\{\mu(1-\rho)T^{(1)}\}^{l-s}}{(l-s)!}\right] = \frac{1}{2^{2l-s}}\binom{2l-s-1}{l-s}$$

Substituting in Equation (5.3), the first part of the result in Lemma 5.1 is obtained. Since $E[R_l]$ is independent of ρ, therefore, this value remains the same when $\rho \to 0^+$, which proves the second part of Lemma 5.1.

Remark 5.1 *Similar to the case of symmetric n-dimensional fork-join queues, it is demonstrated with simulations that the parameter m_n in Equation 5.1 is independent of the sub-task service rate μ. Therefore, for a system with fixed number of parallel tasks n and sub-tasks l, the value of m_n remains unchanged irrespective of changes to service rate μ and job arrival rate λ. In most applications, λ and μ are more likely to change frequently than n and l. Therefore, once $m_{(n,l)}$ is estimated using one simulation, it can be used to estimate the mean response time for any value of λ and μ as long as n and l don't change.*

Remark 5.2 *Conjecture 5.1 is satisfied when the number of tasks $n = 1$. In this case, the system is a tandem $M/M/1$ queueing network. The mean response time in steady state is $\frac{l\rho}{\mu(1-\rho)} + \frac{l}{\mu}$, with $m_{(1,l)} = l$ and $M_{(1,l)} = \frac{l}{\mu}$.*

The performance of Conjecture 5.1 for the case of $n = 5$, $l = 5$, and $\mu = 1$ is demonstrated in Figure 5.2. The linear relationship between the simulated mean response time and $\frac{l\rho}{\mu(1-\rho)}$ is evident from the straight line observed.

Conjecture 5.1 leads to an approximation algorithm for the mean response time of a symmetric tandem fork-join queueing network in steady state, along the same lines as Algorithm 1 in Chapter 4. For completion, this is presented as Algorithm 2. Following the same naming convention, this algorithm is called n-l-1 linear approximation algorithm.

In the rest of this chapter, for brevity, $E[T_{(n,l)}]$ is denoted by $\overline{T}_{(n,l)}$, the estimate using the n-l-1 linear approximation algorithm is denoted by $\widehat{\overline{T}}_{(n,l)}$ and the estimate obtained using simulations is denoted by $\widehat{\overline{T}}_{(n,l)}^{sim}$. The estimate of the parameter $m_{(n,l)}$ in Equation (5.1) is denoted by $\widehat{m}_{(n,l)}$.

The use of the n-l-1 linear approximation algorithm is now demonstrated with a numerical example.

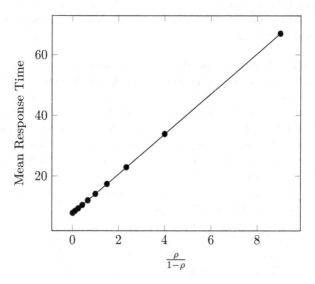

Figure 5.2: Plot of simulated average response time vs. $\frac{\rho}{1-\rho}$ for $n = 5$, $l = 5$, and $\mu = 1$

Algorithm 2 n-l-1 Linear approximation algorithm for computation of $\widehat{\widehat{T}}_{(n,l)}$

Input: Number of parallel tasks, n; Number of sub-tasks, l
Output: Expression for $\widehat{\widehat{T}}_{(n,l)}$ $\forall \rho \in [0,1)$, *i.e.* $\forall \lambda \in [0,\mu)$, $\mu > 0$.

1: Simulate the system for $\rho = 0.5$ and $\mu = 1$. Estimate expected response time $\widehat{\widehat{T}}_{(n,l)}^{sim}(0.5)$.

2: Compute $M_{(n,l)}$

3: Solve this linear equation for $\widehat{m}_{(n,l)}$: $\widehat{\widehat{T}}_{(n,l)}^{sim}(0.5) = \frac{0.5\widehat{m}_{(n,l)}}{\mu(1-0.5)} + M_{(n,l)}$

4: **return** $\widehat{\widehat{T}}_{(n,l)}(\rho) = \frac{\rho\widehat{m}_{(n,l)}}{\mu(1-\rho)} + M_{(n,l)}$

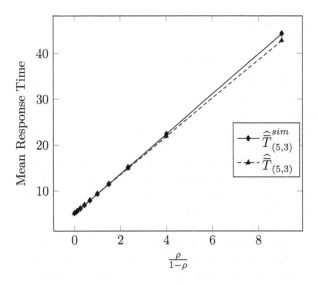

Figure 5.3: Plot of simulated and approximated (using Algorithm 2) mean response times vs. $\frac{\rho}{1-\rho}$ for $n = 5$, $l = 5$, and $\mu = 1$

5.1.2 Numerical Example

Consider a symmetric tandem fork-join queueing network with $n = 5$ and $l = 3$. First, a simulation of this system with $\rho = 0.5$ and $\mu = 1$ is run to obtain the value of $\widehat{\overline{T}}_{(3,5)}^{sim}(0.5)$ to be equal to 9.37 time units. Computing the expectation of the maximum of five independent Erlang-$(3,1)$ random variables, the value of $M_{(5,3)} = 5.19$. Finally, the following linear equation is solved for $\widehat{m}_{(5,3)}$: $9.37 = \frac{0.5\widehat{m}_{(5,3)}}{1(1-0.5)} + 5.19$, to obtain $\widehat{m}_{(5,3)} = 4.17$. Therefore, $\forall \rho \in [0, 1)$ and $\mu \geq 0$, the following approximation is obtained for $\overline{T}_{(3,5)}$: $\widehat{\overline{T}}_{(3,5)} = \frac{4.17\rho}{\mu(1-\rho)} + \frac{5.19}{\mu}$. The performance of this algorithm for this example is demonstrated in Figure 5.3. The efficacy of the approximation is evident from the closeness of the values of mean response times obtained using: (i) simulations, and (ii) the n-l-1 linear approximation algorithm.

5.1.3 Experimental Results

In this section, comparisons between estimates of the mean response time obtained using simulations and the n-l-1 linear approximation algorithm (Algorithm 2) are presented. Results are reported for values of $l = 2, 3, 4$, and 5, and $n = 2, 5$, and 10 (Tables (5.1)–(5.6)). A sub-task service rate of $\mu = 1$ is assumed. In the tables in this section, for specific values of n and l, $\widehat{T}^{sim}_{(n,l)}$ denotes the expected response time estimate obtained using simulations, $\widehat{T}_{(n,l)}$ denotes that obtained using the n-l-1 linear approximation algorithm and % Error denotes the percentage difference between the two estimates. The calculation of the error percentage is similar to the calculation in Equation 4.11. Results are reported for $\rho = 0.1$ to 0.9 with increments of 0.1 excluding $\rho = 0.5$. Results for $\rho = 0.5$ have been excluded since the simulated mean response time estimate for $\rho = 0.5$ is used in the solution of the linear equation in the n-l-1 linear approximation algorithm.

Simulation results are based on results from 10 simulation runs. For each simulation run, 30 million job inter-arrival times and the corresponding n times l service times for each job were generated. The response time of each job was recorded. The response times of the first 10 million jobs were not included in the calculation of the mean to account for the warm-up period. Response times of the last 20 million jobs were used for the calculation of the mean response time in steady state. 95% confidence intervals are reported for each average response time. These confidence intervals were generated using the average response times in the 10 simulation runs.

When $n = 2$ (Tables (5.1)–(5.2)), the error percentages are within 2%, the highest being 1.912% for $l = 5$ and $\rho = 0.9$. When $n = 5$ (Tables (5.3)–(5.4)), the error percentages are within 5% with the maximum being 4.117% for $l = 5$ and $\rho = 0.9$. Lastly, when $n = 10$ (Tables (5.5)–(5.6)), the error percentages are within 6%, the maximum being 5.617% when $l = 5$ and $\rho = 0.9$. The increase in error percentages with increase in ρ can be attributed to the fact that more time is required to approximate steady state accurately in simulations as ρ increases.

The time required for simulations ranged from 37 minutes and 29 seconds to 4 hours, 33 minutes and 45 seconds. In comparison, with the n-l-1 linear approximation algorithm, once the value of $m_{(n,l)}$ is estimated, the average response time for any value of ρ can be estimated in less than a second. The importance of this algorithm is evident from the fact that there are no approximation methods available for estimating the mean response time of the symmetric tandem fork-join queueing network in prior literature.

Table 5.1: Comparison with simulation results for $n = 2$, $l = 2$ and $n = 2$, $l = 3$

ρ	$\widehat{\overline{T}}_{(2,2)}^{sim}$	$\widehat{\overline{T}}_{(2,2)}$	% Error	$\widehat{\overline{T}}_{(2,3)}^{sim}$	$\widehat{\overline{T}}_{(2,3)}$	% Error
0.1	3.02±0.000	3.03	0.077	4.32±0.000	4.33	0.133
0.2	3.37±0.000	3.37	0.112	4.81±0.001	4.82	0.202
0.3	3.81±0.001	3.81	0.115	5.44±0.001	5.45	0.203
0.4	4.40±0.001	4.40	0.080	6.29±0.001	6.30	0.137
0.6	6.48±0.003	6.47	-0.122	9.26±0.003	9.24	-0.212
0.7	8.56±0.005	8.54	-0.284	12.25±0.006	12.19	-0.496
0.8	12.74±0.013	12.68	-0.484	18.25±0.013	18.09	-0.865
0.9	25.26±0.057	25.09	-0.682	36.24±0.049	35.78	-1.284

Table 5.2: Comparison with simulation results for $n = 2$, $l = 4$ and $n = 2$, $l = 5$

ρ	$\widehat{\overline{T}}_{(2,4)}^{sim}$	$\widehat{\overline{T}}_{(2,4)}$	% Error	$\widehat{\overline{T}}_{(2,5)}^{sim}$	$\widehat{\overline{T}}_{(2,5)}$	% Error
0.1	5.59±0.000	5.60	0.183	6.84±0.001	6.85	0.226
0.2	6.22±0.001	6.24	0.276	7.60±0.001	7.63	0.335
0.3	7.03±0.001	7.05	0.278	8.60±0.001	8.63	0.333
0.4	8.13±0.002	8.14	0.189	9.93±0.001	9.96	0.220
0.6	11.99±0.005	11.95	-0.281	14.66±0.004	14.62	-0.331
0.7	15.87±0.010	15.76	-0.662	19.42±0.007	19.27	-0.764
0.8	23.65±0.025	23.38	-1.137	28.96±0.022	28.59	-1.289
0.9	47.03±0.096	46.25	-1.662	57.64±0.078	56.54	-1.912

Table 5.3: Comparison with simulation results for $n = 5$, $l = 2$ and $n = 5$, $l = 3$

ρ	$\widehat{\overline{T}}_{(5,2)}^{sim}$	$\widehat{\overline{T}}_{(5,2)}$	% Error	$\widehat{\overline{T}}_{(5,3)}^{sim}$	$\widehat{\overline{T}}_{(5,3)}$	% Error
0.1	4.14±0.000	4.15	0.211	5.64±0.000	5.66	0.313
0.2	4.56±0.000	4.58	0.327	6.21±0.000	6.24	0.483
0.3	5.11±0.001	5.12	0.339	6.95±0.000	6.99	0.498
0.4	5.84±0.001	5.86	0.234	7.95±0.001	7.98	0.341
0.6	8.45±0.002	8.42	-0.382	11.52±0.002	11.46	-0.535
0.7	11.08±0.005	10.98	-0.918	15.13±0.006	14.94	-1.276
0.8	16.36±0.014	16.09	-1.608	22.39±0.014	21.89	-2.218
0.9	32.25±0.048	31.45	-2.483	44.26±0.064	42.77	-3.383

Table 5.4: Comparison with simulation results for $n = 5$, $l = 4$ and $n = 5$, $l = 5$

ρ	$\widehat{\overline{T}}_{(5,4)}^{sim}$	$\widehat{\overline{T}}_{(5,4)}$	% Error	$\widehat{\overline{T}}_{(5,5)}^{sim}$	$\widehat{\overline{T}}_{(5,5)}$	% Error
0.1	7.07±0.000	7.10	0.395	8.46±0.001	8.50	0.465
0.2	7.78±0.000	7.83	0.610	9.30±0.001	9.37	0.715
0.3	8.71±0.001	8.76	0.627	10.41±0.001	10.49	0.728
0.4	9.97±0.002	10.01	0.426	11.92±0.001	11.98	0.490
0.6	14.46±0.004	14.37	-0.652	17.33±0.003	17.21	-0.732
0.7	19.02±0.007	18.73	-1.521	22.82±0.006	22.43	-1.689
0.8	28.18±0.018	27.45	-2.597	33.84±0.021	32.88	-2.838
0.9	55.76±0.051	53.62	-3.841	66.99±0.099	64.24	-4.117

Table 5.5: Comparison with simulation results for $n = 10$, $l = 2$ and $n = 10$, $l = 3$

ρ	$\widehat{\overline{T}}_{(10,2)}^{sim}$	$\widehat{\overline{T}}_{(10,2)}$	% Error	$\widehat{\overline{T}}_{(10,3)}^{sim}$	$\widehat{\overline{T}}_{(10,3)}$	% Error
0.1	4.99±0.000	5.01	0.314	6.62±0.000	6.65	0.432
0.2	5.47±0.000	5.49	0.493	7.24±0.001	7.29	0.676
0.3	6.08±0.001	6.12	0.518	8.05±0.001	8.11	0.702
0.4	6.92±0.001	6.94	0.361	9.16±0.001	9.21	0.484
0.6	9.90±0.003	9.85	-0.584	13.14±0.004	13.04	-0.769
0.7	12.93±0.005	12.75	-1.413	17.19±0.007	16.88	-1.826
0.8	19.02±0.015	18.55	-2.492	25.35±0.016	24.55	-3.153
0.9	37.39±0.056	35.96	-3.833	49.89±0.084	47.57	-4.659

Table 5.6: Comparison with simulation results for $n = 10$, $l = 4$ and $n = 10$, $l = 5$

ρ	$\widehat{\overline{T}}_{(10,4)}^{sim}$	$\widehat{\overline{T}}_{(10,4)}$	% Error	$\widehat{\overline{T}}_{(10,5)}^{sim}$	$\widehat{\overline{T}}_{(10,5)}$	% Error
0.1	8.15±0.000	8.19	0.532	9.62±0.001	9.68	0.614
0.2	8.91±0.001	8.98	0.827	10.52±0.001	10.62	0.950
0.3	9.91±0.001	10.00	0.856	11.70±0.001	11.82	0.972
0.4	11.28±0.002	11.35	0.586	13.33±0.002	13.42	0.657
0.6	16.23±0.005	16.08	-0.905	19.21±0.004	19.02	-0.994
0.7	21.26±0.009	20.81	-2.103	25.20±0.009	24.63	-2.292
0.8	31.39±0.017	30.28	-3.569	37.27±0.022	35.83	-3.857
0.9	61.94±0.079	58.67	-5.273	73.59±0.118	69.46	-5.617

5.2 HETEROGENEOUS FORK-JOIN QUEUES

Heterogeneous fork-join queues are queueing systems in which the symmetry aspect of the symmetric n-dimensional fork-join queues is relaxed. Symmetry in fork-join queues requires all the task service times to be stochastically identical. In the heterogeneous fork-join queues analyzed in this section, this requirement is relaxed, and the task service times are allowed to take different mean values. A formal problem definition follows in the next section.

5.2.1 System Definition

Jobs arrive into the fork-join queueing system following a Poisson process with rate λ. On arrival, each job splits instantaneously into n tasks. The tasks are processed at n single server queueing stations operating under an FCFS service discipline. The service times of the tasks are independent across jobs and across tasks belonging to the same job. However, these tasks are not identically distributed. Task i, $i = 1, \ldots, n$ is exponentially distributed with rate μ_i. A job is considered complete and leaves the system when all the n tasks are complete. It follows from the work of Baccelli et al [7] that this system is stable *iff* $\lambda < \min_{i=1,\ldots,n} \mu_i$. In the next section, an approximation algorithm to estimate the mean response time of this system in steady state is presented. No other approximations are available in prior literature for this system when $n > 2$.

5.2.2 Response Time Estimation

In the heterogeneous fork-join queueing system under consideration, each task response time is exponential with rate $\mu_i - \lambda$. For given values of λ and μ_i, $i = 1, \ldots, n$, the expectation of the maximum of i *independent* exponentially distributed random variables with rate $\mu_i - \lambda$ is denoted by $M_i(\lambda)$. The response time in steady state is denoted by S_n. The following conjecture forms the basis for estimation of the mean response time in steady state.

Conjecture 5.2 *The mean response time of a heterogeneous fork-join queueing system in steady state, with exponential inter-arrival and service time distributions is given by:*

$$E[S_n] = \sum_{i=1}^{n} (M_i(\lambda) - M_{i-1}(\lambda))(\mu_i - \lambda) \left[\frac{\lambda m_n^{(i)}}{\mu_i(\mu_i - \lambda)} + \frac{1}{\mu_i} \right] \quad (5.4)$$

where, $m_n^{(i)}$ are parameters independent of λ, $m_n^{(1)} = 1$ and $M_0(\lambda) = 0$, $\forall \lambda \in [0, \min_{i=1,\ldots,n} \mu_i)$.

Remark 5.3 *As an example, for $n = 2$, $(M_2(\lambda) - M_1(\lambda))(\mu_2 - \lambda) = \frac{\mu_1 - \lambda}{\mu_1 + \mu_2 - 2\lambda}$. Substituting in Equation 5.4 :*

$$E[S_2] = \frac{1}{\mu_1 - \lambda} + \frac{\mu_1 - \lambda}{\mu_1 + \mu_2 - 2\lambda} \left[\frac{\lambda m_2^{(2)}}{\mu_2(\mu_2 - \lambda)} + \frac{1}{\mu_2} \right]. \qquad (5.5)$$

Substituting $\mu_1 = \mu_2 = \mu$ in Equation (5.5) results in the expression for the mean response time of the symmetric 2-dimensional fork-join queue given by Nelson and Tantawi [45], with $m_2^{(2)} = \frac{3}{4}$. On the other hand, if $\frac{1}{\mu_1 - \lambda} >> \frac{1}{\mu_2 - \lambda}$, $E[S_2] \approx \frac{1}{\mu_1 - \lambda}$. This is true because: if the service rate of one task is much higher than that of the other task, the response time will be dominated by the contribution of the slower task. This domination becomes more effective as the difference between the response times of the two queues increases with increase in the traffic intensity.

Intuition behind Conjecture 5.2. Consider the symmetric fork-join queueing system of Chapter 4 with $n = 2$ in steady state. Once the task with index 1 of any job finishes service, if it is the last of the two tasks to be complete, it departs from the system. If not, it waits in a join buffer for the other task to be finish service. If the waiting time in the join buffer is denoted by R, based on the exact result for $n = 2$ by Nelson and Tantawi [45]:

$$E[R] = \frac{3\lambda}{8\mu(\mu - \lambda)} + \frac{1}{2\mu}. \qquad (5.6)$$

The mean synchronization time is the sum of two components. The first is the mean service time of the second task. Due to the memorylessness property of exponential distribution, the expectation of this is equal to $\frac{M_2(0) - M_1(0)}{\mu_2}$. The other component is the workload in front of the second task in the queue when the first finishes. It was shown by Nelson and Tantawi [45] for $n = 2$ and observed in Chapter 3 for $n > 2$ that this residual expected workload is directly proportional to the average steady-state workload in the second queue, *i.e.* $\frac{\lambda}{\mu(\mu - \lambda)}$ and the proportionality constant is independent of λ. This property is extended to the heterogeneous case to obtain Conjecture 5.2.

Conjecture 5.2 can be used to define the heterogeneous n-1 approximation algorithm to estimate the mean response time in steady state. In this case, there are $n-1$ parameters that need to be estimated. Therefore, $n-1$ simulations for $n-1$ values of the arrival intensity λ are needed. Using results from these $n-1$ simulations, $n-1$ linear equations are solved to estimate the values of the parameters in Conjecture 5.2. These parameters can then be used to estimate the mean response time for all other arrival intensities. This algorithm is formally defined in Algorithm 3. For brevity, $E[S_n]$ is denoted by $\bar{S}_n$, the estimate using the heterogeneous n-1 approximation algorithm is denoted by $\widehat{\bar{S}}_n$ and the estimate obtained using simulations is denoted by $\widehat{\bar{S}}_n^{sim}$. The estimate of the parameter m_n in Equation (5.1) is denoted by $\widehat{m}_n$.

Algorithm 3 Heterogeneous n-1 approximation algorithm for computation of $\widehat{\bar{S}}_n$

Input: Number of parallel tasks, n; Arrival rate λ, Service rates μ_i, $i = 1, \ldots, n$

Output: Expression for $\widehat{\bar{S}}_n \ \forall \lambda \in [0, \min_{i=1,\ldots,n} \mu_i)$

1: Choose any $n-1$ values of $\lambda \in [0, \min_{i=1,\ldots,n} \mu_i)$, $\hat{\lambda}_j$, $j = 2, \ldots, n$. Simulate and estimate expected response times $\widehat{\bar{S}}_n^{sim}(\hat{\lambda}_j)$.

2: Compute $M_i(\hat{\lambda}_j)$, $\forall i = 1, \ldots, n$ and $j = 2, \ldots, n$.

3: Solve the system of linear equations for $\widehat{m}_n^{(i)}$, $i = 1, \ldots, n$:
$\widehat{m}_n^{(1)} = 1$; $\widehat{\bar{S}}_n^{sim}(\hat{\lambda}_j) = \sum_{i=1}^{n}(M_i(\hat{\lambda}_j) - M_{i-1}(\hat{\lambda}_j))(\mu_i - \hat{\lambda}_j)\left[\frac{\hat{\lambda}_j \widehat{m}_n^{(i)}}{\mu_i(\mu_i - \hat{\lambda}_j)} + \frac{1}{\mu_i}\right]$, $\forall j = 2, \ldots, n$.

4: **return** $\widehat{\bar{S}}_n(\lambda) = \sum_{i=1}^{n}(M_i(\lambda) - M_{i-1}(\lambda))(\mu_i - \lambda)\left[\frac{\lambda \widehat{m}_n^{(i)}}{\mu_i(\mu_i - \lambda)} + \frac{1}{\mu_i}\right]$

Remark 5.4 *Consider a heterogeneous fork-join queueing system with n tasks, each with service rate μ_i, $i = 1, \ldots, n$. If a network controlling authority is considering adding a new task with service rate μ_{n+1} while keeping the initial network the same, the structure of Algorithm 3 is such that the estimated values of $m_n^{(i)}$ will remain the same in the new network with $n+1$ tasks. Only one simulation will be needed to estimate $m_{n+1}^{(n+1)}$.*

5.2.3 Numerical Example and Results

The use of the heterogeneous n-1 approximation algorithm (Algorithm 3) is now demonstrated with a numerical example. Consider the heterogeneous fork-join queueing system with $n = 2$, $\mu_1 = 1$ and $\mu_2 = 1.5$. Conjecture 5.2 for this case takes the form in Equation (5.5). Simulations are run for $\hat{\lambda}_2 = 0.5$ to obtain $\widehat{\overline{S}}_2^{sim}(0.5) = 2.216$ time units. $M_1(0.5) = \frac{1}{1-0.5} = 2$ and $M_2(0.5) = 2.333$. The resulting linear equation is then solved to get $\widehat{m}_2^{(2)} = 0.322$. Finally, this is substituted in Equation (5.5) to obtain approximations for the mean response time.

The performance of the heterogeneous n-1 approximation algorithm is demonstrated by comparing the estimates of the mean response time obtained from simulations and the heterogeneous n-1 approximation algorithm. In this section, the results are reported while varying values of the arrival intensity λ as opposed to the traffic intensity ρ in previous sections since the traffic intensity is different for each task node. Simulations follow similar specifications as described in Section 5.1.3. $\widehat{\overline{S}}_n^{sim}$ denotes the value of the expected steady-state response times obtained using simulations. These are reported along with their 95% confidence intervals. $\widehat{\overline{S}}_n$ denotes the estimate obtained using the heterogeneous n-1 approximation algorithm. The error percentages were calculated as in Equation 4.11. Results are reported for arrival intensities corresponding to $\lambda = 0.05, \ldots, 0.9$ at intervals of 0.05.

In the first system (Table 5.7), the parameters are the same as that in the numerical example. Only one simulation is required in this case. Simulations for $\lambda = 0.5$ were used to estimate the single unknown parameter. The error percentages are encouraging, with the maximum being 0.454% when the arrival intensity is 0.9.

In the second system (Table 5.8), $n = 3$, $\mu_1 = 0.5$, $\mu_2 = 1.0$ and $\mu_3 = 1.5$. In this case, it would have been possible to use the value of the unknown parameter obtained in the previous system with $n = 2$ as suggested in Remark 5.4. However, the results reported here are calculated by following Algorithm 3 explicitly. Simulations for arrival intensities of 0.150 and 0.350 are used to estimate the two unknown parameters. In this case too, the error percentages are less than 1%, with the maximum being 0.802% when the arrival intensity is 0.450.

So far in this chapter, promising results have been demonstrated for the symmetric tandem fork-join queueing network and

Table 5.7: Comparison with simulation results for $n = 2$, $\mu_1 = 1.0$, and $\mu_2 = 1.5$

λ	$\widehat{\overline{H}}_2^{sim}$	$\widehat{\overline{H}}_2$	% Error
0.05	1.32±0.000	1.32	-0.049
0.10	1.38±0.000	1.38	-0.094
0.15	1.45±0.000	1.45	-0.128
0.20	1.52±0.000	1.52	-0.151
0.25	1.61±0.000	1.61	-0.160
0.30	1.71±0.000	1.70	-0.157
0.35	1.82±0.000	1.81	-0.139
0.40	1.94±0.000	1.94	-0.107
0.45	2.10±0.001	2.09	-0.059
0.55	2.49±0.001	2.50	0.072
0.60	2.77±0.001	2.77	0.149
0.65	3.12±0.001	3.12	0.230
0.70	3.58±0.002	3.59	0.312
0.75	4.23±0.003	4.25	0.381
0.80	5.21±0.006	5.23	0.427
0.85	6.84±0.012	6.87	0.454
0.90	10.12±0.031	10.16	0.420

Table 5.8: Comparison with simulation results for $n = 3$, $\mu_1 = 0.5$, $\mu_2 = 1.0$, and $\mu_3 = 1.5$

λ	$\widehat{\overline{H}}_3^{sim}$	$\widehat{\overline{H}}_3$	% Error
0.025	2.53±0.000	2.53	0.013
0.050	2.64±0.000	2.65	0.016
0.075	2.77±0.001	2.77	0.016
0.100	2.91±0.001	2.91	0.012
0.125	3.07±0.001	3.07	0.006
0.175	3.46±0.001	3.46	-0.009
0.200	3.70±0.001	3.70	-0.019
0.225	3.99±0.002	3.99	-0.029
0.250	4.34±0.002	4.34	-0.036
0.275	4.77±0.002	4.77	-0.042
0.300	5.30±0.003	5.30	-0.044
0.325	6.00±0.004	5.99	-0.033
0.375	8.23±0.007	8.23	0.067
0.400	10.19±0.012	10.22	0.198
0.425	13.49±0.021	13.55	0.438
0.450	20.11±0.049	20.27	0.802

the heterogeneous fork-join queueing system. In the next section, a preliminary analysis is presented for a form of (n,k) fork-join queues with promising applications.

5.3 (n, k) FORK-JOIN QUEUES

In analyzing the (n, k) fork-join queueing system, the restriction on the number of tasks that need to finish service before the job is considered complete is relaxed. Joshi et al [29] first introduced these (n, k) fork-join queues that are encountered in network coding algorithms. Arrivals into the system are according to a Poisson process with rate λ. As in previous sections, the job is split into n tasks that are processed in parallel at n queueing stations operating based on an FCFS service discipline. Each task requires independent service times distributed exponentially with rate μ. So far, the system description is identical to that in Chapter 4. However, in (n, k) fork-join queues, out of the n tasks, only k need to finish service for the job to be considered complete. At the time instant when the service of a task i, $1 \leq i \leq n$, is about to start, if k other tasks belonging to the same job are already complete, the service time of task i is zero. However, if k tasks are not complete, task i begins its service and remains in service until its service completion or the completion of k tasks of its parent job, whichever happens first.

Joshi et al [29] show that this (n, k) fork-join queueing system is stable *iff* $\lambda < \frac{n\mu}{k}$. The authors compute bounds for the response time of the system. This remains the only work on (n, k) fork-join queues under the above assumptions.

In this system, the workload at a single task station is denoted by $T_{(n,k)}^{(i)}$, $1 \leq i \leq n$. Due to symmetry, $E[T_{(n,k)}^{(1)}] = E[T_{(n,k)}^{(2)}] = \ldots = E[T_{(n,k)}^{(n)}]$. The steady-state response time random variable is denoted by $T_{(n,k)}$. The following is a conjecture on the relationship between $E[T_{(n,k)}]$ and $E[T_{(n,k)}^{(1)}]$:

Conjecture 5.3 *In steady state, the mean response time of an (n, k) fork-join queueing system with exponential inter-arrival and service time distributions is linear with respect to the mean response time of a single task. The relationship between the two quantities can be expressed as:*

$$E[T_{(n,k)}] = m_{(n,k)} \left(E[T^{(1)}] - \frac{k}{n\mu} \right) + \frac{H_n - H_{k-2}}{\mu} \qquad (5.7)$$

where, $m_{(n,k)}$ is a parameter independent of ρ and H_n is the nth harmonic sum.

The term $\frac{H_n - H_{k-2}}{\mu}$ in Equation (5.7) is the mean response time of a job that enters an empty system, i.e. the expected value of the k-th order statistic of n i.i.d. exponentially distributed random variables with rate μ. Since Equation (5.7) is written in terms of the mean response time of a single queue and not the mean workload in a single queue, the average service time is subtracted from the expected response time.

Remark 5.5 In the variations of fork-join queues analyzed previously, the realizations of the service times at any task were independent of those at other tasks. In the (n,k) fork-join queueing system, this is no longer the case. Smaller service times at the first k tasks to finish service will result in smaller service times at the remaining $n - k$ tasks too. This opens up the possibility that the idea behind the conjectures in Chapters 4 and 5 could be extended to systems in which the task service times are not independent.

5.3.1 Comparison with Simulations

Preliminary simulation results are now presented that support Conjecture 5.3. In Figure 5.4, the simulated mean response time of the system in steady state is plotted on the y-axis and the mean time spent by a job at only one task, denoted by $E[T_{(n,k)}^{(1)}]$ is plotted on the x-axis. The straight line supports Conjecture 5.3.

Unfortunately, unlike the previous variations of fork-join queues analyzed, Conjecture 5.3 does not directly lead to an expression for $E[T_{(n,k)}^{(1)}]$. This is because the individual task queues do not behave as independent systems. In a given queue, the service time of a task might decrease depending on the service times of other tasks belonging to the same parent job. Since the dependence structure of this system is complex, estimation of the expected response time of a single task is harder than other fork-join queues analyzed previously. However, efficient bounds on the individual task response times can be used to obtain bounds on the response time of the system using Conjecture 5.3. This opens up prospects for future work.

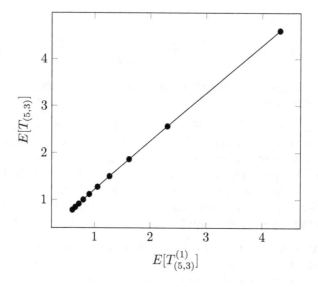

Figure 5.4: Plot of $E[T_{(5,3)}^{(1)}]$ vs. $E[T_{(5,3)}]$

5.4 CONCLUSIONS

In this chapter, simple and easy to compute estimates are proposed for the mean response time of three variants of fork-join queues. These queueing networks are so hard to analyze that no approximations for any of the performance measures of these systems exist in prior literature. The proposed approximations are based on conjectures that are strongly supported by simulations. The analyses presented here will be useful for making decisions on system design in real-world applications.

For example, in a symmetric tandem fork-join queueing network, division of one task into sub-tasks and having a separate queueing station for each sub-task might result in an improvement in the average response time. However, if there is a cost (labor cost, for example) for operating each queueing station, then the trade-off between the benefit gained from lower response times and the drawback of increased cost needs to be evaluated. The analysis presented in this chapter enables such trade-off evaluations without requiring the decision-maker to make any potentially expensive physical changes.

Similarly, in the case of heterogeneous fork-join queues, the heterogeneous n-1 approximation algorithm can be used to estimate the change in mean response time with an increase or decrease in the number of parallel tasks. This saves computing resources and the time required to run simulations for every possible value of arrival and service rates.

Finally, in the (n, k) fork-join queueing system, if the server starts processing a specific task, and k other tasks finish service before the service completion of that task, then the server has wasted some amount of time and resources in serving a task that did not require service. Therefore, it is not obvious that the response time is minimized by sending the job to all n task queues and waiting for k of them to finish. For example, simulations show that for lower traffic intensities, when $(n, k) = (5, 3)$, the system performs better in terms of lower expected response times than when $(n, k) = (10, 6)$. However, for higher traffic intensities, the system with $(n, k) = (10, 6)$ performs better for the same value of the service rate μ. Therefore, bounds that can be computed in future using Conjecture 5.3 can be used to construct an efficient system specific to the application.

This chapter concludes the discussion on fork-join queueing networks in this book.

Bibliography

[1] Surprise! First Dual-Core Smartphone Arrives Early. https://gigaom.com/2010/12/16/surprise-first-dual-core-smartphone-arrives-early/, 2010.

[2] M. H. Ammar and S. B. Gershwin. Equivalence relations in queueing models of fork/join networks with blocking. *Performance Evaluation*, 10(3):233–245, 1989.

[3] F. Baccelli. Two parallel queues created by arrivals with two demands:the M/G/2 symmetrical case. Technical Report RR-0426, INRIA, 1985.

[4] F. Baccelli and P. Bremaud. *Palm Probabilities and Stationary Queues*. Springer Science & Business Media, 2012.

[5] F. Baccelli and Z. Liu. On the execution of parallel programs on multiprocessor systems–a queuing theory approach. *Journal of the ACM*, 37(2):373–414, April 1990.

[6] F. Baccelli and A. Makowski. Simple computable bounds for the fork-join queue. In *Proceedings of the Conference of Information Science Systems, John Hopkins University*, 436–441, 1985.

[7] F. Baccelli, A. M. Makowski, and A. Shwartz. The fork-join queue and related systems with synchronization constraints: stochastic ordering and computable bounds. *Advances in Applied Probability*, 21(3):629–660, 1989.

[8] F. Baccelli and W. A. Massey. Series-parallel fork-join queueing networks and their stochastic ordering. Technical Report RR-0534, INRIA, 1986.

[9] F. Baccelli, W. A. Massey, and D. Towsley. Acyclic fork-join queuing networks. *Journal of the ACM*, 36(3):615–642, 1989.

[10] S. Balsamo, L. Donatiello, and N. M. Van Dijk. Bound performance models of heterogeneous parallel processing systems. *IEEE Transactions on Parallel and Distributed Systems*, 9(10):1041–1056, 1998.

[11] S. Balsamo and I. Mura. Approximate response time distribution in fork and join systems. *ACM SIGMETRICS Performance Evaluation Review*, 23(1):305–306, 1995.

[12] S. Balsamo and I. Mura. On queue length moments in fork and join queuing networks with general service times. In *Computer Performance Evaluation Modelling Techniques and Tools*, 218–231, 1997.

[13] S. Banerjee, P. Gupta, and S. Shakkottai. Towards a queueing-based framework for in-network function computation. *Queueing Systems*, 72(3-4):219–250, 2012.

[14] O. Boxma, G. Koole, and Z. Liu. Queueing-theoretic solution methods for models of parallel and distributed systems. In *Performance Evaluation of Parallel and Distributed Systems Solution Methods. CWI Tract 105 & 106*, 1996.

[15] L. Breiman. Random Forests. *Machine Learning*, 45(1):5–32, 2001.

[16] N. Carmeli, G. B. Yom-Tov, and O. J. Boxma. State-dependent estimation of delay distributions in fork-join networks. *Eurandom Preprint Series*, 2018.

[17] P. M. Chen, E. K. Lee, G. A. Gibson, R. H. Katz, and D. A. Patterson. RAID: High-performance, reliable secondary storage. *ACM Computing Surveys*, 26(2):145–185, 1994.

[18] R. J. Chen, H. Zhang, and H. Hu. A fast simulation for thousands of general homogeneous fork/join queues. In *International Conference on Intelligent Systems, Modelling and Simulation*, 300–305, 2010.

[19] IBM Corporation. Power 4: The first multi-core, 1GHz processor. http://www-03.ibm.com/ibm/history/ibm100/us/en/icons/power4/.

[20] H. Dai. Exact monte carlo simulation for fork-join networks. *Advances in Applied Probability*, 43(2):484–503, 2011.

[21] Y. Dallery, Z. Liu, and D. Towsley. Equivalence, reversibility, symmetry and concavity properties in fork-join queuing networks with blocking. *Journal of the ACM*, 41(5):903–942, 1994.

[22] A. Duda and T. Czachórski. Performance evaluation of fork and join synchronization primitives. *Acta Informatica*, 24(5):525–553, 1987.

[23] L. Flatto. Two parallel queues created by arrivals with two demands II. *SIAM Journal on Applied Mathematics*, 45(5):861–878, 1985.

[24] L. Flatto and S. Hahn. Two parallel queues created by arrivals with two demands I. *SIAM Journal on Applied Mathematics*, 44(5):1041–1053, 1984.

[25] N. Gautam. *Analysis of Queues: Methods and Applications*. CRC Press, Inc., USA, 1st edition, 2012.

[26] S. B. Gershwin. Assembly/disassembly systems: An efficient decomposition algorithm for tree-structured networks. *IIE Transactions*, 23(4):302–314, 1991.

[27] Singapore Government. Lamppost as a platform. https://www.tech.gov.sg/scewc2019/laap.

[28] V. D. R. Guide Jr., G. C. Souza, and E. Van Der Laan. Performance of static priority rules for shared facilities in a remanufacturing shop with disassembly and reassembly. *European Journal of Operational Research*, 164(2):341–353, 2005.

[29] G. Joshi, Y. Liu, and E. Soljanin. On the delay-storage tradeoff in content download from coded distributed storage systems. *IEEE Journal on Selected Areas in Communications*, 32(5):989–997, 2014.

[30] G. K. David. Stochastic Processes Occurring in the theory of queues and their analysis by the method of the imbedded Markov chain. *The Annals of Mathematical Statistics*, 24(3):338–354, 1953.

[31] B. Kemper and M. Mandjes. Mean sojourn times in two-queue fork-join systems: bounds and approximations. *OR spectrum*, 34(3):723–742, 2012.

[32] R. S. Khabarov, V. A. Lokhvitckii, and A. S. Dudkin. Relationship invariants based sojourn time approximation for the fork-join queueing system. In *Models and Methods of Information Systems Research Workshop in the frame of the Betancourt International Engineering Forum (MMISR 2019)*, 63–68, 2019.

[33] S. Ko and R. F. Serfozo. Response times in M/M/s fork-join networks. *Advances in Applied Probability*, 36(3): 854–871, 2004.

[34] S. Ko and R. F. Serfozo. Sojourn times in G/M/1 fork-join networks. *Naval Research Logistics (NRL)*, 55(5):432–443, 2008.

[35] A. Krishnamurthy, R. Suri, and M. Vernon. Analysis of a fork/join synchronization station with inputs from coxian servers in a closed queuing network. *Annals of Operations Research*, 125(1):69–94, 2004.

[36] A. Kumar and R. Shorey. Performance analysis and scheduling of stochastic fork-join jobs in a multicomputer system. *IEEE Transactions on Parallel and Distributed Systems*, 4(10):1147–1164, 1993.

[37] S. S. Lavenberg. A perspective on queueing models of computer performance. *Performance Evaluation*, 10(1):53–76, 1989.

[38] A. S. Lebrecht and W. J. Knottenbelt. Response time approximations in fork-join queues. In 23rd UK Performance Engineering Workshop (UKPEW), 2007.

[39] E. K. Lee and R. H. Katz. An analytic performance model of disk arrays. *ACM SIGMETRICS Performance Evaluation Review*, 21(1):98–109, 1993.

[40] H. Li and S. H. Xu. On the dependence structure and bounds of correlated parallel queues and their applications to synchronized stochastic systems. *Journal of Applied Probability*, 37(4):1020–1043, 2000.

[41] Y. Liu. Queueing network modeling of elementary mental processes. *Psychological Review*, 103(1):116–136, 1996.

[42] Y. C. Liu and H. G. Perros. A decomposition procedure for the analysis of a closed fork/join queueing system. *IEEE Transactions on Computers*, 40(03):365–370, 1991.

[43] J. C. S. Lui, R. R. Muntz, and D. Towsley. Computing performance bounds of fork-join parallel programs under a multiprocessing environment. *IEEE Transactions on Parallel and Distributed Systems*, 9(3):295–311, 1998.

[44] J. Menon and D. Mattson. Performance of disk arrays in transaction processing environments. In *Proceedings of the 12th International Conference on Distributed Computing Systems*, 302–309, 1992.

[45] R. Nelson and A. N. Tantawi. Approximate analysis of fork/join synchronization in parallel queues. *IEEE Transactions on Computers*, 37(6):739–743, 1988.

[46] R. Nelson, D. Towsley, and A. N. Tantawi. Performance analysis of parallel processing systems. *IEEE Transactions on Software Engineering*, 14(4):532–540, 1988.

[47] M. Nguyen, S. Alesawi, N. Li, H. Che, and H. Jiang. A blackbox fork-join latency prediction model for data-intensive applications. *IEEE Transactions on Parallel and Distributed Systems*, 31(9):1983–2000, 2020.

[48] V. Nguyen. Processing networks with parallel and sequential tasks: Heavy traffic analysis and brownian limits. *The Annals of Applied Probability*, 3(1):28–55, 1993.

[49] F. Pedregosa, G. Varoquaux, A. Gramfort, V. Michel, B. Thirion, O. Grisel, M. Blondel, P. Prettenhofer, R. Weiss, V. Dubourg, J. Vanderplas, A. Passos, D. Cournapeau, M. Brucher, M. Perrot, and E. Duchesnay. Scikit-learn: Machine learning in Python. *Journal of Machine Learning Research*, 12:2825–2830, 2011.

[50] D. Pinotsi and M. A. Zazanis. Synchronized queues with deterministic arrivals. *Operations Research Letters*, 33(6):560–566, 2005.

[51] Z. Qiu, J. F. Pérez, and P. G. Harrison. Beyond the mean in fork-join queues: Efficient approximation for response-time tails. *Performance Evaluation*, 91:99–116, 2015.

[52] M. I. Reiman and B. Simon. Open queueing systems in light traffic. *Mathematics of Operations Research*, 14(1):26–59, 1989.

[53] A. Rizk, F. Poloczek, and F. Ciucu. Computable bounds in fork-join queueing systems. In *Proceedings of the 2015 ACM SIGMETRICS International Conference on Measurement and Modeling of Computer Systems*, 335–346, 2015.

[54] M. S. Squillante, Y. Zhang, A. Sivasubramaniam, and N. Gautam. Generalized parallel-server fork-join queues with dynamic task scheduling. *Annals of Operations Research*, 160(1):227–255, 2008.

[55] M. Takahashi, H. Osawa, and T. Fujisawa. On a synchronization queue with two finite buffers. *Queueing Systems*, 36(1):107–123, 2000.

[56] M. Takahashi and Y. Takahashi. Synchronization queue with two map inputs and finite buffers. In *Proceedings of the Third International Conference on Matrix Analytical Methods in Stochastic Models, Leuven, Belgium*, 2000.

[57] SEE Lab Technion. Ed patient flow. https://www. youtube. com/watch?v=HP_au996Ffw.

[58] A. Thomasian. Analysis of fork/join and related queueing systems. *ACM Computing Surveys*, 47(2):1-71, 2014.

[59] A. Thomasian and J. Menon. RAID5 performance with distributed sparing. *IEEE Transactions on Parallel and Distributed Systems*, 8(6):640–657, 1997.

[60] A. Thomasian and A. N. Tantawi. Approximate solutions for M/G/1 fork/join synchronization. In *Proceedings of the 26th Conference on Winter Simulation*, 361–368, 1994.

[61] D. Towsley, C. G. Rommel, and J. A. Stankovic. Analysis of fork-join program response times on multiprocessors. *IEEE Transactions on Parallel and Distributed Systems*, 1(3):286–303, 1990.

[62] D. Towsley and S. Yu. Bounds for two server fork-join queueing systems. *Computer and Information Science [COINS]*, 1987.

[63] I. Tsimashenka and W. J. Knottenbelt. Reduction of subtask dispersion in fork-join systems. In *European Workshop on Performance Engineering*, 325–336, 2013.

[64] E. Varki. Mean value technique for closed fork-join networks. *ACM SIGMETRICS Performance Evaluation Review*, 27: 103–112, 1999.

[65] E. Varki. Response time analysis of parallel computer and storage systems. *IEEE Transactions on Parallel and Distributed Systems*, 12(11):1146–1161, 2001.

[66] E. Varki and L. W. Dowdy. Response time analysis of two server fork-join systems. In *Proceedings of MASCOTS '96 - 4th International Workshop on Modeling, Analysis and Simulation of Computer and Telecommunication Systems*, 291–295, 1996.

[67] E. Varki, A. Merchant, and H. Chen. The M/M/1 fork-join queue with variable sub-tasks. *unpublished, available online*.

[68] E. Varki and C. Zhang. Quick performance bounds for computer and storage systems with parallel resources. *unpublished, available online*.

[69] S. Varma. Heavy and light traffic approximations for queues with synchronization constraints. *PhD Thesis*, 1990.

[70] S. Varma and A. M. Makowski. Interpolation approximations for symmetric fork-join queues. *Performance Evaluation*, 20(1–3):245–265, 1994.

[71] P. E. Wright. Two parallel processors with coupled inputs. *Advances in Applied Probability*, 24(4):986–1007, 1992.

[72] S. Wu, S. Jiang, B. C. Ooi, and K. Tan. Distributed online aggregations. In *Proceedings of the VLDB Endowment*, 2: 443–454, 2009.

[73] C. H. Xia, Z. Liu, D. Towsley, and M. Lelarge. Scalability of fork/join queueing networks with blocking. *ACM SIGMETRICS Performance Evaluation Review* 35(1):133–144, June 2007.

[74] Y. Zeng, A. Chaintreau, D. Towsley, and C. H. Xia. Throughput scalability analysis of fork-join queueing networks. *Operations Research*, 66(6):1728–1743, 2018.

[75] H. Zhao, C. H. Xia, Z. Liu, and D. Towsley. Distributed resource allocation for synchronous fork and join processing networks. In *2010 Proceedings IEEE INFOCOM*, 1–5, 2010.

[76] H. Zhao, C. H. Xia, Z. Liu, and D. Towsley. A unified modeling framework for distributed resource allocation of general fork and join processing networks. *ACM SIGMETRICS Performance Evaluation Review*, 38(1):299–310, 2010.

Index